Waste Management in Spatial Environments

The increasing scarcity of land and the ever-rising amount of waste produced worldwide, coupled with the consequent change of focus by policy makers from waste disposal and recovery to waste prevention, is boosting research in the 'economics of waste'.

This volume addresses waste-management and waste-disposal issues, embedding them in spatial, systemic and trade-related frameworks. The collection is policy oriented, including socio-economic and political science perspectives in order to provide an understanding of real-world phenomena, and thus maximize its value for policy making.

The book includes contributions on the linkages between income and waste generation and landfilling (such as the 'waste Kuznets curve' conceptual framework), in addition to chapters that bring together policy-oriented analysis of instrument effectiveness and the spatial nature of waste phenomena. On top of this, there are pieces of research emphasizing technological spillovers and trade at interregional and intercountry levels. The comparative analysis of policy effectiveness and efficiency at the regional and country levels is also covered, including the assessment of the potential role of illegal management of waste in determining waste performance. To give a spatial and comparative flavour, the book includes examples covering household, industrial and special waste.

The wide set of methodologies and issues included in this book make it a comprehensive starting point for scholars and policy makers interested in waste-related research.

Alessio D'Amato is Assistant Professor and Lecturer in Public Economics and Environmental Economics at the University of Rome 'Tor Vergata', Italy. His research activity is focused on the theory of incentives, environmental regulation under asymmetric information, waste policy in the presence of illegal disposal and organized crime, and emissions trading.

Massimiliano Mazzanti is Associate Professor in the Department of Economics and Management at the University of Ferrara, Italy. His research deals with environmental policy, the economics of innovation, waste management and policy, and climate change and development.

Anna Montini is Assistant Professor in Economics and Lecturer in Environmental Economics at the University of Bologna, Italy. Her main research interests lie in environmental economics and policy, waste management and environmental-economic performance at the spatial level.

Routledge Studies in Ecological Economics

1. Sustainability Networks
Cognitive tools for expert collaboration in social-ecological systems
Janne Hukkinen

2. Drivers of Environmental Change in Uplands
Aletta Bonn, Tim Allot, Klaus Hubaceck and Jon Stewart

3. Resilience, Reciprocity and Ecological Economics
Northwest coast sustainability
Ronald L. Trosper

4. Environment and Employment
A reconciliation
Philip Lawn

5. Philosophical Basics of Ecology and Economy
Malte Faber and Reiner Manstetten

6. Carbon Responsibility and Embodied Emissions
Theory and measurement
João F.D. Rodrigues, Alexandra P.S. Marques and Tiago M.D. Domingos

7. Environmental Social Accounting Matrices
Theory and applications
Pablo Martínez de Anguita and John E. Wagner

8. Greening the Economy
Integrating economics and ecology to make effective change
Bob Williams

9. Sustainable Development
Capabilities, needs, and well-being
Edited by Felix Rauschmayer, Ines Omann and Johannes Frühmann

10. The Planet in 2050
The Lund discourse of the future
Edited by Jill Jäger and Sarah Cornell

11. From Bioeconomics to Degrowth
Georgescu-Roegen's 'New Economics' in eight essays
Edited by Mauro Bonaiuti

12. Socioeconomic and Environmental Impacts on Agriculture in the New Europe
Post-Communist transition and accession to the European Union
S. Serban Scrieciu

13. Waste and Recycling
Theory and empirics
Takayoshi Shinkuma and Shusuke Managi

14. Global Ecology and Unequal Exchange
Fetishism in a zero-sum world
Alf Hornborg

15. The Metabolic Pattern of Societies
Where economists fall short
Mario Giampietro, Kozo Mayumi and Alevgül H. Sorman

16. Energy Security for the EU in the 21st Century
Markets, geopolitics and corridors
Edited by José María Marín-Quemada, Javier García-Verdugo and Gonzalo Escribano

17. Hybrid Economic-Environmental Accounts
Edited by Valeria Costantini, Massimiliano Mazzanti and Anna Montini

18. Ecology and Power
Struggles over land and material resources in the past, present and future
Edited by Alf Hornborg, Brett Clark and Kenneth Hermele

19. Economic Theory and Sustainable Development
What can we preserve for future generations?
Vincent Martinet

20. Paving the Road to Sustainable Transport
Governance and innovation in low-carbon vehicles
Edited by Måns Nilsson, Karl Hillman, Annika Rickne and Thomas Magnusson

21. Creating a Sustainable Economy
An institutional and evolutionary approach to environmental policy
Edited by Gerardo Marletto

22. The Economics of Climate Change and the Change of Climate in Economics
Kevin Maréchal

23. Environmental Finance and Development
Sanja Tišma, Ana Maria Boromisa and Ana Pavičić Kaselj

24. Beyond Reductionism
A passion for interdisciplinarity
Edited by Katharine Farrell, Tommaso Luzzati and Sybille van den Hove

25. The Business Case for Sustainable Finance
Edited by Iveta Cherneva

26. The Economic Value of Landscapes
Edited by C. Martijn van der Heide and Wim Heijman

27. Post-Kyoto Climate Governance
Confronting the politics of scale, ideology and knowledge
Asim Zia

28. Climate Economics
The state of the art
Frank Ackerman and Elizabeth A. Stanton

29. Good Governance, Scale and Power
A case study of North Sea fisheries
Liza Griffin

30. Waste Management in Spatial Environments
Edited by Alessio D'Amato, Massimiliano Mazzanti and Anna Montini

Waste Management in Spatial Environments

Edited by

Alessio D'Amato, Massimiliano Mazzanti
and
Anna Montini

Routledge
Taylor & Francis Group
LONDON AND NEW YORK

First published 2013
by Routledge
2 Park Square, Milton Park, Abingdon, Oxon OX14 4RN

Simultaneously published in the USA and Canada
by Routledge
605 Third Avenue, New York, NY 10017

Routledge is an imprint of the Taylor & Francis Group, an informa business

British Library Cataloguing in Publication Data
A catalogue record for this book is available from the British Library

Library of Congress Cataloging in Publication Data
Waste management in spatial environments / edited by Alessio D'Amato, Massimiliano Mazzanti and Anna Montini.
pages cm
Includes bibliographical references and index.
1. Refuse and refuse disposal–Italy–Case studies. 2. Refuse disposal industry–Italy–Management. 3. Commercial policy–Environmental aspects. 4. International trade. I. D'Amato, Alessio. II. Mazzanti, Massimiliano. III. Montini, Anna.
HD4485.I83W37 2013
363.72'80945–dc23
2012044635

ISBN: 978-0-415-68718-8 (hbk)
ISBN: 978-0-203-38322-3 (ebk)

Typeset in Times New Roman by
Out of House Publishing

Contents

Figures

Tables

Contributors

Graziano Abrate is Assistant Professor of Management at the University of Eastern Piedmont and Research Fellow at HERMES. His research interests include the fields of management science, industrial organization and public sector management. His current work focuses on productivity and demand-side management issues in the provision of local public services (waste, water, collective transport, electricity). His research has been published in international peer-reviewed journals including *Regional Studies*, *Regional Science and Urban Economics*, *Journal of Productivity Analysis*, *Transportation Research Part E*, *IEEE Transactions on Power Systems* and *Tourism Management*.

Dario Biolcati Rinaldi is a postgraduate student in the Faculty of Economics at the University of Ferrara. His main research interests include environmental economics with a particular focus on the behaviour of agents, the multi-tasking model, warm-glow giving and 'impure altruism'.

Alessandro Bucciol is Assistant Professor of Econometrics at the University of Verona and a Netspar affiliate. He held postdoctoral research positions in the universities of Padua and Amsterdam. His research interests are in households' savings and portfolio decisions, risk analysis, behavioural economics and welfare evaluation of pension policies. He has published on these topics in several academic journals including the *Review of Economics and Statistics*, *Journal of the European Economic Association*, *Journal of Economic Psychology* and *Journal of Economic Behavior and Organization*.

Giacomo D'Alisa graduated in 2004 after studying the Economics of International Trade and Market Value at 'Parthenope' University of Naples (Italy); in 2006 he completed a higher training scheme in Management of Sustainable Development and a higher education course in the History of Ethics and Political Thought. He was awarded a PhD in Economics and Technologies for Sustainable Development from the University of Foggia (Italy) in May 2010 and took up a position as visiting researcher at the

Institute of Science and Environmental Technology of the Autonomous University Barcelona in Spain until becoming a research fellow at that institution. His research interests include environmental economics, ecological economics, sustainable development, public goods, commons, social metabolism and environmental conflicts.

Alessio D'Amato is Assistant Professor and Lecturer in Public Economics and Environmental Economics at the University of Rome 'Tor Vergata', where he was awarded his PhD in Economics (Teoria Economica ed Istituzioni). He is affiliated to SEEDS (Sustainability, Environmental Economics and Dynamics Studies). His research activity is focused on: theory of incentives; environmental regulation under asymmetric information; waste policy in the presence of illegal disposal and organized crime; and emissions trading.

Simon De Jaeger is Assistant Professor at the Center for Research on Economic Markets and their Environments, HU Brussel, Belgium. He is an affiliated researcher at the Center for Economic Studies, Faculty of Business and Economics, KU Leuven, Belgium. His research interests include the economics of municipal solid waste, local policy evaluation, spatial feedback effects and spillovers of policy instruments, efficiency measurement and environmental Kuznets curves in waste management.

Maria Federica Di Nola holds a bachelor's degree in Economics from the University of Naples 'Federico II' and a master's degree in Economics from the University of the Basque Country, UPV/EHU. She is currently studying for a PhD at UPV/EHU in collaboration with the University of Rome 'La Sapienza'. Her research focuses on the use of System Dynamics to analyse the waste management crisis in the Campania region (southern Italy) that began in 1990.

Matteo Ferraris graduated from the University of Torino (Italy), and was awarded a master's degree in Economics from Coripe Piemonte (Collegio Carlo Alberto, Moncalieri) and a PhD ('Decisions in Insurance and Finance') from the University of Torino. He is a research fellow at CNR-Ceris in Milan and collaborates with the ITC-ILO Centre at the University of Turin as a tutor on their masters' programme. His areas of research are economics of public utilities, public economics and environmental economics with a particular focus on waste management and labour markets, derivatives and corporate finance, and public procurement.

Derek Kellenberg is Associate Professor in the Department of Economics at the University of Montana, and Faculty Associate at the Montana Institute on Ecosystems. He has published articles on international pollution havens, international waste trade and the economics of natural disasters in the *Journal of Environmental Economics and Management*, the

Journal of International Economics and the *Journal of Urban Economics*, among others.

Massimiliano Mazzanti is Associate Professor in Economics and Lecturer in Environmental Economics at the University of Ferrara. He is a research collaborator with the CERIS-DSE CNR Institute in Milan and a member of SEEDS (Sustainability, Environmental Economics and Dynamics Studies). His research focuses on environmental policy, economics of innovation, economic performance and innovation, economic evaluation by stated preference techniques, waste management and policy, climate change and development.

Natalia Montinari is Postdoctoral Research Fellow at the Max Planck Institute of Economics in Jena. In 2011 she was awarded a PhD in Economics from the University of Padua. Her research interests are in waste economics, behavioural and experimental economics with focus on contract theory, and the evolution of other-regarding preferences in children.

Anna Montini is Assistant Professor in Economics and Lecturer in Environmental Economics at the University of Bologna. She is a Research Fellow at the National Research Council in Milan (CERIS-DSE) and a member of SEEDS (Sustainability, Environmental Economics and Dynamics Studies). Her main research interests are environmental economics and policy, waste management and environmental-economic performance at the geographical/spatial level.

Francesco Nicolli is a Postdoctoral Research Fellow at DEIT, Department of Economics, University of Ferrara and is affiliated to SEEDS (Sustainability, Environmental Economics and Dynamic Studies). His main research interests hinge on econometric analysis of environmental policy and technological change. He was awarded a doctorate in Applied Economics from the University of Ferrara, and received a MSc in Environment and Resource Economics from the University of Birmingham, UK.

Marco Piovesan is Associate Professor at FOI, University of Copenhagen. Before joining FOI, he was a CLER Fellow at the Harvard Business School and Assistant Research Professor in the Department of Economics at the University of Copenhagen. His work has been published in several academic journals including the *American Economic Review*, *Economics Letters*, *PLoS ONE*, the *Journal of Economic Behavior and Organization* and the *Journal of Economic Psychology*.

Virginia Turchi is a postgraduate student in the Faculty of Economics at the University of Ferrara. Her research interest is environmental and development economics with a specific focus on policy evaluation and policy effectiveness.

Lorenzo Valmasoni graduated from the University of Padua after studying the effectiveness of monetary incentives in waste sorting. Following an internship at the Economic Research Division of the Italian Central Bank, he enrolled as a PhD student in the area of Experimental Economics at the University of Erlangen-Nurenberg.

Michela Zappaterra is a postgraduate student in the Faculty of Economics at the University of Ferrara. Her research interest is waste economics and environmental policy.

Roberto Zoboli is Full Professor of Economic Policy at the Catholic University of Milan, Italy, and Director of the Master in International Relations in the same university. For many years he has been Research Director at CERIS, the Institute of Research on Firms and Growth of the National Research Council of Italy, and he has coordinated various research projects on environmental policy under the umbrella of the European Environment Agency.

Mariangela Zoli is Assistant Professor in Economic Policy and Lecturer in Public Economics and Environmental Economics at the University of Rome 'Tor Vergata'. She is a member of SEEDS (Sustainability, Environmental Economics and Dynamics Studies). Her research activity is mainly focused on waste policy and illegal behaviours; emissions trading taxation; sustainable consumption; the environmental impact of redistributive policies; optimal income taxation; and multidimensional poverty.

Abbreviations and acronyms

Arpav	Veneto Regional Environmental Protection Agency
COMEXT	intra- and extra-EU trade data
COMTRADE	Common Format for Transient Data Exchange
CONAI	Consorzio Nazionale Imballaggi
COPERT	Computer Programme to calculate Emissions from Road Transport
DG TREN	DG Transport and Energy
DHA	Density of Human Activity
DtD	door-to-door
DWD	Density of Waste Disposed
EEA	European Environment Agency
EIONET	European Environment Information and Observation Network
EKC	Environmental Kuznets Curve
EMS	Environmental Management Systems
E-PRTR	The European Pollutant Release and Transfer Register
ETC/SCP	European Topic Centre Sustainable Consumption and Production
EU	European Union
ExternE	External Costs of Energy
FAO	Food Agricultural Organization
FDI	Foreign Direct Investment
FE	Fixed Effects
GCR	Global Competitiveness Reports
GDP	Gross Domestic Product
GFD	gate fee (negative price) in destination country
GFH	gate fee (negative price) at a hypothetical site in home country
GHG	greenhouse gases
GNI	Gross National Investments
ICH	intermediate input cost for collection in the home country origin of the waste

ICTH	intermediate input costs of transport services in the home country
IMD	impact of management at destination
IMH	impact of equivalent management in home country
ISPRA	Istituto Superiore per la Protezione e la Ricerca Ambientale (Institute for Environmental Protection and Research)
ISTAT	Istituto nazionale di statistica (Italian National Institute of Statistics)
ITD	impact of transport at destination abroad
ITH	impact of hypothetical transport in home country
IWT	international waste trade
JRC	Joint Research Centre
LCA	life cycle analysis
MSW	municipal solid waste
MuSIASEM	Multi-Scale Integrated Analysis of Societal and Ecosystem Metabolism
NI	net total impact
NMI	net environmental impact of IWT in management
NS	Not Separated Waste
NTI	net impact of IWT in transport
NVAD	net value added for treatment in the destination country
NVAH	net value added for treatment in the home country
OECD	Organisation for Economic Co-operation and Development
OLS	Ordinary Least Squares
PAYT	pay-as-you-throw
PCSW	Per Capita Sorted Waste
PCTW	Per Capita Total Waste
PD	price of the material/energy from waste in the destination country
PH	price of the material/energy from waste in the home country
PHD	price of the waste input in the destination country
PHW	price of the waste input in the home country
PPML	Poisson Pseudo Maximum Likelihood
PRGRU	Piano Regionale per la Gestione dei Rifiuti Urbani
PWD	price for waste as an input in the country of destination
PWH	price of waste in the domestic treatment industry
RE	Random Effects
RES	Renewable Energy Sources
SLAB	Spent Lead Acid Batteries
SWR	Sorted Waste Ratio
TARES	Tassa sui rifiuti e servizi

TARSU	Tariffa Ambientale sui Rifiuti Solidi Urbani
TCD	transportation cost for delivery to the foreign treatment facility
TCH	transportation cost for delivery to the home treatment facility
TD	transport cost in destination country
TH	transport cost for hypothetical management at home
TIA	Tariffa Integrata Ambientale
TOT	Total Waste
UK	United Kingdom
US	United States
VA	value added
WCR	World Competitiveness Report
WD	Waste Disposed
WDMR	Waste Disposed Metabolic Rate
WG	Waste Generation
WGMR	Waste Generated Metabolic Rate
WKC	Waste Kuznets Curve
WS	Waste Separated

Introduction

Alessio D'Amato, Massimiliano Mazzanti and Anna Montini

The 'economics of waste' is a field of environmental economics which features a growing interest in theoretical analyses dealing with the design of optimal policy packages in first-best and second-best situations, and in the presence of non-competitive market scenarios, illegal market rents and dynamic issues. The increasing scarcity of land and the consequential change in policy, focused first on waste disposal and recovery and then on waste production, have also generated a need for empirical analyses providing evidence on policy effectiveness, the relevance of specific regional features and the impact of a comprehensive set of socio-economic drivers. Numerous microeconomic oriented studies and macroeconomic analyses, in streams linked to the Environmental Kuznets Curve (EKC) literature, have emerged during the past years, broadening the 'management' focus to include wider economic considerations, and complementing the historically stronger literature on the valuation of waste related externalities, with the aim of offering robust food for thought for central and decentralized policy makers.

This book is a collection of theoretical and empirical chapters addressing waste management and waste disposal issues and embedding them in spatial, systemic and trade related frameworks. The collection is policy oriented. It includes heterodox economic, socio-economic and political sciences perspectives in order to increase the value for policy making, to provide an understanding of real world phenomena and to increase the book's readership. Although some contributions contain applications from the economics toolkit, the approach is always heterodox in kind, and diversified in its methodology. Complementarities between techniques and views on waste management issues are sought.

Empirical contributions provide new insights and evidence regarding achievement of full decoupling in waste generation and landfilling of waste (e.g. the 'Waste Kuznets Curve' conceptual framework), which are specific core targets of current and future European Union (EU) policies. We collect chapters that bring together policy oriented analysis of instrument effectiveness and the spatial nature of waste phenomena, with emphasis on technological spillovers and trade/shipments at inter-regional and inter-country levels. The volume includes comparative analysis of policy effectiveness and efficiency, focusing on the role of instruments such as unit based pricing and landfill taxes, at the regional and country levels, as well as studies that touch on the role of illegal management of waste in determining waste performance. Studies cover household, industrial and special waste.

We describe below the conceptual frameworks and original value provided by this book and its focus on two macro themes: analysis of waste decoupling embedded in spatial environments, including convergence analyses and policy assessment in highly decentralized and heterogeneous regional settings; analyses of the drivers of waste shipments at intra- or inter-country level. The contributions of the book are based on streams of (new) research in the waste realm.

The EU's 'thematic strategies' on resources and waste include reference to indicators of 'absolute' and 'relative' decoupling (European Commission, 2003a, 2003b; Jacobsen *et al.*, 2004), the former being the negative relationship between economic growth and environmental impact and the latter being the positive but decreasing-in-size relationship. The thematic strategies theoretically refer to the EKC conceptual framework. The new EU 2020 strategy for sustainable growth and jobs also recognizes the central role of 'resource efficiency' (extended to waste) and decoupling as pillars for the construction of a greener and more competitive (knowledge) economy (EU documents 'Boost Resource Efficiency and Recycling' and '2020: A New Economic Strategy').

The achievement of some degree of decoupling is of prime importance given that evidence for the EU (EEA, 2009) shows an absence of even relative decoupling for waste generation. The European Environment Agency (EEA, 2009) acknowledges that waste volumes in the EU are growing, driven by changing production and consumption patterns (Andersen *et al.*, 2007), and highlights (EEA, 2006) the importance of flexible implementation of market-based instruments, to be managed within a decentralized approach to environmental policy in the EU, to achieve higher levels of decoupling. Policy effectiveness and spatial phenomena are interrelated and are important for achieving waste targets through effective (diffusion of) waste policies in intrinsically decentralized frameworks, in terms of policy implementation and management. Within a varied policy framework, convergence or divergence of the decoupling performance, among as well as within countries, must be compared as part of the effort to achieve overall targets based on effectiveness and efficiency criteria. Policy decentralization and idiosyncratic socio-economic elements can generate very different dynamics even under the same (EU) regulatory umbrella.

Analyses of decoupling trends and the effects of local structural and waste management factors on waste related performance are valuable in examining (1) spatial dimensions and (2) convergence in waste performance between the leaders and laggards (for example, average northern and southern areas of Italy). These analyses are rooted in the assessment of waste generation and landfilled waste drivers, which provide socio-economic, structural and (endogenous) waste management instruments. We show that achieving decoupling should not be taken for granted since economic growth and reduction in the efficiency of resources' use could quickly reverse the situation. The analysis of waste performance drivers, spatial phenomena and convergence are complementary: policy makers and waste managers need to know what is driving performance and which among the significant drivers could be most easily influenced by waste actors. It is necessary also to know whether waste management in provinces is independent or is part of the activities of a cluster (intra-regional, inter-regional). The degree of convergence tells us whether the laggards are catching up. Ultimately, it is average national performance that matters. We can consider drivers and spatial-based knowledge as part of a puzzle that we need to complete to improve the worst performances and eventually homogenize trends.

Italy is a useful case because it has high levels of decentralization in environmental policy making (much higher than in the UK) similar to the levels in Spain and Germany, and is moving towards an even stronger federal organization. Also, Italy is characterized by major income differences and, historically, quite different economic and environmental performance in the northern and southern regions. Divergence or convergence in current and future waste performance is an important issue that is receiving renewed attention since the (both practical and financial) collapse of the waste management schemes in Naples, Palermo and some other areas in 2008–2009. These are important issues given that countries are monitored and valued on the basis of their national average performance and also that regional system collapses have to be financed by national taxes. The findings in the case of Italy can be extended to policy and waste management schemes in other highly decentralized and heterogeneous environments.

A specific issue that has been overlooked, partly due to data unavailability, is Transboundary Waste Shipments (TWS). There are two main policy-relevant questions. First, there are economic as well as *environmental impacts* associated with the TWS that occur in Europe. Although for industries in the home and destination countries economic impacts, e.g. employment, may be important, mapping and quantifying the environmental impacts of TWS is crucial from a policy perspective. These impacts may have direct implications for current evaluations of the legal and policy regimes for TWS in Europe (and worldwide), which are mainly aimed at minimizing environmentally harmful consequences. Environmental impact measurements of TWS do not explore the decision processes behind the international flows.

Second, understanding TWS *drivers* requires careful analysis of economic and institutional factors, including environmental regulation in different countries, to explain why and how TWS take place. This analysis could have important policy implications: by highlighting the decision mechanisms leading to TWS it might suggest possible levers for European and national policies to control, reduce or influence TWS.

The policy relevance of this issue is linked to the possibility that a better understanding of TWS could improve the environment. We investigate the factors behind the decisions taken by waste producing/collecting (supply) and waste treatment (demand) enterprises. In general, leaving aside illegal behaviors, their decision to ship internationally or to manage domestically are driven by economic and technological considerations, and we expect that they may lead to *cost-saving decisions.*

In a recent contribution on waste trading, Kellenberg (2012) notes that the structure of waste flows encompasses both waste traded before any decision on recycling is taken and waste that is traded after recycling has occurred. Differences in economic, institutional, structural and regulatory country based factors may explain waste trade. Kellenberg (2012) stresses the importance of structural (population density), market price (gate fees), and technology/capacity factors, as well as regulatory stringency and enforcement. Cost/price and technology/capacity factors, along the lines of the debate on the 'pollution haven hypothesis', can drive trade in waste in the same direction or induce a balance. In general, lower (management/disposal) prices and higher capital intensity (i.e. incineration, recycling sites) – reflecting economies of scale and comparative advantages in recycling and disposal – should attract waste flows. In the context of trade issues, 'relative' rather than absolute factors are the more relevant. At an empirical level, Baggs (2009) finds that capital intensity attracts imports of hazardous waste. Kellenberg (2012), using international COMTRADE data for 92 countries and bilateral trade flows on both hazardous and non-hazardous waste, finds that waste imports increase for countries with less strict environmental regulation vis-à-vis the trading partner, implying a sort of 'waste haven hypothesis'. Within-country level analysis could help our understanding of international flows. De Jaeger and Eyckmans (2010) find for Flanders municipalities that for some waste – bulky household refuse, demolition waste, garden waste – the quantities collected at the local recycling center depend on the prices charged at recycling centers in neighboring municipalities. Within-country trade can be influenced by the degree of spatial correlation in waste performance across regions in the country. The evidence is highly country-specific and depends mostly on institutional and economic factors influencing the structure of waste regulations from the center to the periphery. Ham (2009) finds some degree of convergence in recycling for UK municipalities and robust spatial correlations (local areas behave similarly). At a higher administrative level (Italian provinces), Mazzanti *et al.* (2009 and 2010) find some signs of convergence in waste trends, but also find negligible and decreasing spatial correlations over time. Policy decentralization, also linked to differentiated economic development, can generate situations where waste patterns differ and where waste management specialization among regions and states differs. In general, we can say that the more uncorrelated and diverging the waste trends are at the intra-country and international levels, the more likely it is that waste trade will emerge. Heterogeneity drives trade and potential win-win market exchanges. Policies should regulate trade and avoid the creation of critical hotspots, particularly in densely populated areas.

Future research based on the contributions in this book could examine the role of economic, legal and technological factors that might have relevance for trade decisions. Some of these drivers are highlighted in EEA (2009).

The main references in the recent literature along these lines of reasoning are Kellenberg (2012), De Jaeger and Eyckmans (2010), Pasotti (2010a, b), Van Beukering and Bouman (2001), Mazzanti *et al.* (2010, 2012). The reader might also be interested in the special issue of *Environmental Economics and Policy Studies*, edited by Alessio D'Amato, Shunsuke Managi and Massimiliano Mazzanti (2012).

References

Andersen, F., Larsen, H., Skovgaard, M., Moll, S. and Isoard, S. (2007), A European Model for Waste and Material Flows. *Resources, Conservation and Recycling*, 49:421–35.

Baggs, J. (2009), International Trade in Hazardous Waste. *Review of International Economics*, 17:1–16.

D'Amato, A., Managi, S. and Mazzanti, M. (2012), Economics of Waste Management and Disposal: Decoupling, Policy Enforcement and Spatial Factors. *Environmental Economics and Policy Studies*, 14:323–5.

De Jaeger, S. and Eyckmans, J. (2010), Do Households Export their Recyclable Waste? Paper presented at the world wcERE conference, Montreal, June–July 2010.

Ham, Y.J. (2009), Convergence of Recycling Rates in the UK: A Spatial Econometrics Perspective. Paper presented at the annual EAERE conference, Amsterdam, 24–26 June.

EEA (2006), *Market-Based Instruments for Environmental Policy in Europe (Technical Report No 8/2005)*. Copenhagen: European Environment Agency.

EEA (2009), *Diverting Waste from Landfills*. Copenhagen: European Environment Agency.

European Commission (2003a), *Towards a Thematic Strategy for Waste Prevention and Recycling*, COM (2003) 301. Brussels: European Commission.

European Commission (2003b), *Towards a Thematic Strategy on Sustainable Use of Natural Resources*, COM (2003) 572. Brussels : European Commission.

Jacobsen, H., Mazzanti, M., Moll, S., Simeone, M.G., Pontoglio, S. and Zoboli, R. (2004), Methodology and Indicators to Measure Decoupling, Resource Efficiency, and Waste Prevention. ETC/WMF, European Topic Centre on Waste and Material Flows, European Environment Agency, P6.2–2004. Copenhagen.

Kellenberg, D. (2012), Trading Wastes. *Journal of Environmental Economics and Management*, 64:68–87.

Mazzanti, M., Montini, A. and Nicolli, F. (2009), Embedding Landfill Diversion in Economic, Geographical and Policy Settings: Regional and Provincial Evidence From Italy, in M. Mazzanti and A. Montini (eds), *Waste & Environmental Policy*, London: Routledge, pp. 126–53.

Mazzanti, M., Montini, A. and Nicolli, F. (2010), Waste Generation and Landfill Diversion Dynamics: Decentralized Management and Spatial Effects. Nota di lavoro 27, FEEM, Milan.

Mazzanti, M., Montini, A. and Nicolli, F. (2012), Waste Dynamics in Economic and Policy Transitions: Decoupling, Convergence and Spatial Effects. *Journal of Environmental Planning and Management*, 55:563–581.

Pasotti, E. (2010a), *Political Branding in Cities: The Decline of Machine Politics in Bogota, Naples, and Chicago.* Cambridge: Cambridge University Press.

Pasotti, E. (2010b), Sorting through the Trash: The Waste Management Crisis in Southern Italy. *South European Society & Politics*, 15:289–307.

Van Beukering, P. and Bouman, M. (2001), Empirical Evidence on Recycling and Trade of Paper and Lead in Developed and Developing Countries. *World Development*, 29:1717–37.

Part I

The Italian environment of waste management

Spatial analyses, convergence, illegal markets and policy assessments

1 A dynamic assessment of Italian landfill taxes

Dario Biolcati Rinaldi, Francesco Nicolli, Virginia Turchi and Michela Zappaterra

Introduction

European environmental policies are aimed primarily at reducing the amount of waste that goes to landfill. The effectiveness of these policies in managing the generation and disposal of waste depends on the efficiency of their implementation.

Efforts to reduce landfill are a priority in the European waste hierarchy, and one of the pillars of the EU's waste strategy is the 1999 Landfill Directive (EEA, 2009), which is being implemented at member state level in association with national efforts to manage waste, such as separated collection, recycling, incineration, and disposal and handling of waste. Most actions are devoted to diverting waste from landfill, and reducing waste generated at source, to achieve a decoupling of different stages of the waste production chain. The European Environment Agency (EEA) recognizes the importance of these actions as volumes of waste in the EU grow, driven by changing production and consumption patterns.

Indicators of decoupling are used frequently to detect and measure improvements in environmental effects and resources efficiency with respect to economic activity. The Organisation for Economic Co-operation and Development (OECD, 2003, 2002) has carried out extensive studies on decoupling to establish indicators for reporting and policy-evaluation purposes. The EU's 'thematic strategies' for resources and waste include references to 'absolute' and 'relative' indicators of delinking (Jacobsen *et al.*, 2004). The first refers to a negative relationship between economic growth and environmental pressures; the second refers to a positive but decreasing-in-size association, which in economic terms is a positive, lower than unity elasticity. Absolute and relative delinking trends are embedded in the more general Environmental Kuznets Curve (EKC) framework (Stern, 2004).

In the EU, landfill is still the predominant solution for the disposing of the municipal waste. In Italy, waste disposal is constantly monitored and evaluated and, in 2007, about 46.7 per cent of total municipal waste went to landfill and 10.3 per cent was incinerated. However, dependence on landfill varies significantly across countries (Mazzanti and Zoboli, 2009; Mazzanti *et al.*, 2008, 2011, 2012; Mazzanti and Montini, 2009; Mazzanti and Nicolli, 2011 provide analyses of waste generation, landfilling and recycling in the EU and Italy). D'Amato *et al.* (2011) analyse how levels of legal disposal (landfill), illegal disposal and recycling in Italy are influenced by tariffs on waste and by crime. Such idiosyncratic factors are relevant in decentralized environments.

Economic analysis of landfilling activity focuses predominantly on cost–benefit assessments of relative externalities. The Institute for Environmental Studies (IVM, 2005) report on the effectiveness of landfill taxes in the EU is a rare exception. Attempts have been made to evaluate the EU Landfill Directive and the now well-established (since 1996) UK landfill tax, but given the lack of extensive (panel) data these are only qualitative assessments. A study by Morris *et al.* (1998) on the early implementation of the UK landfill tax offers some insights on potential and expected contributions to sustainable waste management. The authors analyse its general structure, comparative landfill costs and the waste hierarchy. Morris and Read (2001) and Burnley (2001) provide updates of this analysis, which highlight some operational weaknesses. Burnley (2001) links the EU directive to national implementation in the UK. A rather pessimistic assessment is provided by Martin and Scott (2003), who stress that the imposition of a tax has not changed the behaviour of domestic waste producers to any significant degree. The UK landfill tax was intended to be an incentive to recover, recycle, re-use and minimize waste. Martin and Scott find evidence of more recycling, but not related to re-use or waste minimization. Davies and Doble (2004) monitor the effects of the UK landfill tax from its introduction and provide some forecasts of its evolution, criticalities and externalities. Phillips *et al.* (2007) provide one of the most recent UK-specific regional assessments of waste strategies. Regional analyses, though rare, are useful since the implementation of environmental and resources/waste taxes is often decentralized and associated with costs and benefits.

This chapter analyses delinking trends related to landfilling of municipal solid waste (MSW), within a framework that includes geographic and policy variables.

We focus specifically on the effects of landfill taxes on landfill diversion, in order to assess whether the level of the tax, which might also be capturing elements linked to 'policy commitment', affects landfill performance.

The Italian case is relevant because its environmental policy making is highly decentralized, and has not been accurately analysed. One major constraint is often the lack of reliable data that the decentralization process generates as side effects, in absence of sound central coordination. This impedes the implementation of economic assessments. For the analysis in this chapter, we created an original panel dataset (1999–2008 for the 20 Italian regions) that merges economic and environmental information from ISTAT and ISPRA/APAT.[1]

Analysis of Italy's landfill tax is important at the international level. It was implemented in 1996 before the UK tax (Martin and Scott, 2003; EEA, 2009; Pearce, 2004; DEFRA, 2005). The UK tax was defined and is administered by Her Majesty's Treasury; in Italy landfill taxes (as well as many other competencies) are defined by and are the responsibility of the 20 Italian regions. This decentralization of competencies has, in many fields, including environmental issues, increased since the reform under Article 5 of the Italian Constitution. Taxation and tax revenues are managed by the regions under the general guidelines provided by the Italian Treasury.

The landfill tax is the main environmental tax in Italy and generated around €185 million in revenue in 2010. This amount has decreased consistently over time from a peak of €360 million in 1997. It represents around 38 per cent of total tax revenue (circa half a billion euros from environmental and resources taxation in Italy and 0.005 per cent of total environmental and energy tax revenues, see Figures 1.1–1.3). The reduction in tax revenue is related to the decrease in landfill. However, what is of interest is whether this change in waste disposal is due (even in part) to the imposition of the tax.

To overcome the lack of official data we surveyed regional implementation of the tax based on information provided on official websites, complemented by telephone interviews with regional offices to fill gaps and confirm the information gleaned from the web pages. The result was a reasonably full panel dataset on which to base our econometric analysis. To our knowledge, this is the first analysis to use a long panel dataset (econometric assessments based on cross-section analyses suffer from several problems) on landfill tax implementation.

The research hypothesis is then the following. We want to shed light on whether levels of landfill tax and its dynamics are significant drivers of landfill diversion, that is, whether landfill tax is a relevant omitted variable in past studies on landfill diversion.

Ancillary hypotheses are tested. First, whether the landfill taxes implemented in contiguous regions play a role in determining the focal region's landfilling performance. Disposing of waste is costly, but we cannot rule out the presence of spatial effects associated with government actions (Brueckner, 2003). A very high landfill tax may become an incentive for landfilling outside the regional boundary – especially for provinces close to regional borders. We test also for whether 'technological relative capacity', captured by relative incinerated waste per capita, affects levels of landfilling in the region. Lack of technology may drive landfilling or create incentives for exploiting facilities in nearby regions.

Table 1.1 presents the research hypotheses.

The rest of the chapter is organized as follows. The first section describes the dataset and the empirical model used to assess landfill tax effects; the second section presents the main results; the last section concludes the chapter.

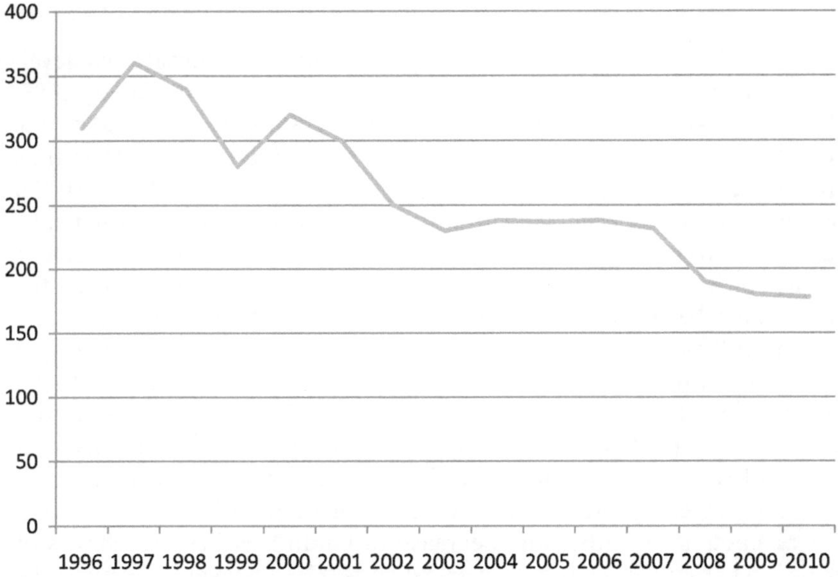

Figure 1.1 Landfill tax revenue 1996–2010 (millions €).

Source: ISTAT Rome.

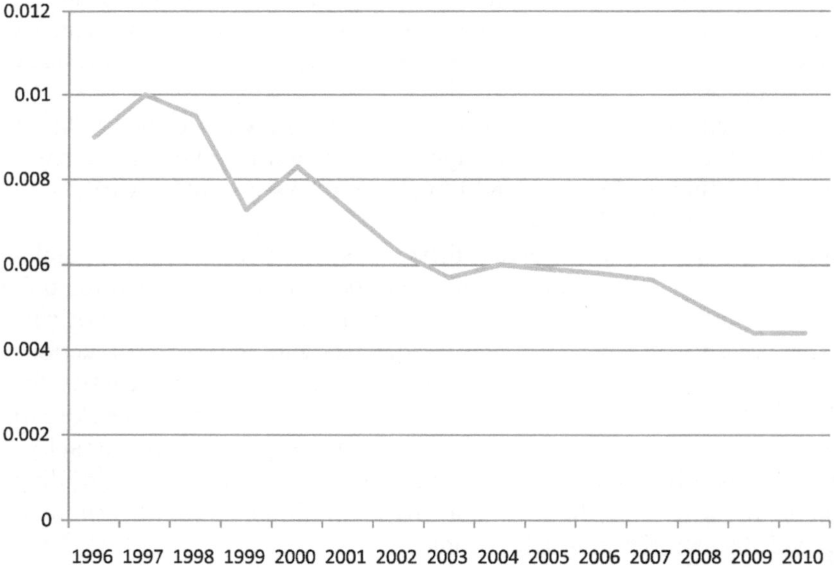

Figure 1.2 Landfill tax revenue as share of total energy-environmental taxation 1996–2010.

Source: ISTAT Rome.

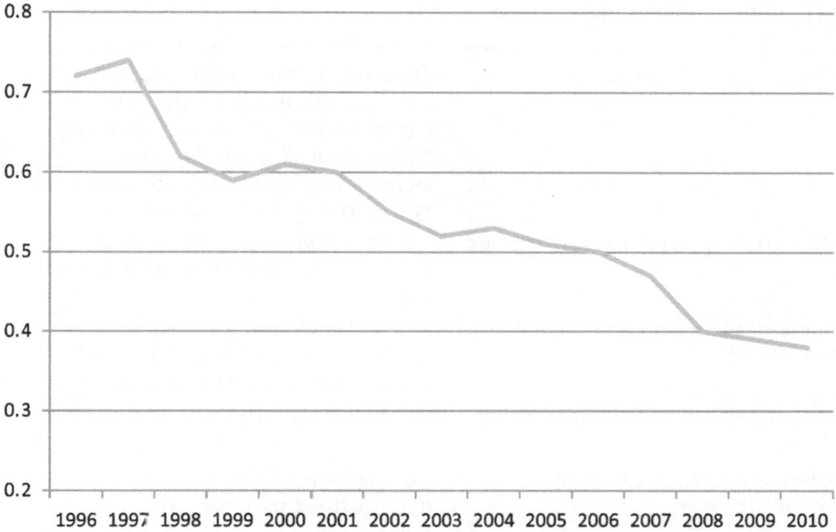

Figure 1.3 Landfill tax revenue as share of total environmental and resource taxation 1996–2010.

Source: ISTAT Rome.

The empirical model and the data

Landfill tax data

Data on regional landfill taxes in Italy were collected through the regional laws determining amounts of tax. Regions were required to implement landfill taxes under national Law 549/1995; however, the timing of their introduction varied across regions. Most fulfilled the requirements of the national law to impose the new tax within 12 months. However, it took seven years for Valle d'Aosta, Molise and Puglia to implement regional laws. Amendments to the national law referred to landfill tax adoption, the definition of waste, and the distribution of responsibilities among different regional offices. To investigate changes in the levels of taxation during the period considered, all regional laws enacted from 1999 to 2008 were analysed. To clarify different definitions of waste streams across municipalities, and to check our interpretation of regional regulations, we conducted telephone interviews with representatives of the regional authorities from January to March 2012. This enabled more homogeneous and comparable definitions of the different waste streams, and consequently more reliable landfill tax data.

Table 1.2 presents levels of landfill taxes from 1999 (the first year for which waste data are available in Italy) to 2008.[2] The dynamics show stable trends, which are in line with other results for environmental and resources taxation. There were few adjustments since implementation, which means that

Table 1.1 Main research hypotheses

Landfill taxation in the region	*Higher/increasing taxes negatively correlate to landfilled waste per capita since they incentivize dynamic reallocation of disposal towards incineration and possibly more recovery of materials in waste management.*
Landfill taxation in contiguous regions	*The higher the value, the more likely waste is not 'exported' to other nearby regions.*
Incinerated waste per capita in the region	*The higher the value, the more likely waste is not landfilled due to technological installed capacity.*
Share of separated collection of waste	*The higher the value, the more likely waste is not landfilled in the coming year.*
Incinerated waste per capita in contiguous regions	*The higher the value, the more likely waste disposal is 'exported' to other nearby regions due to nearby technological installed capacity.*

Note: all variables used for testing implications show cross-section and time related variation.

taxes are subject to an erosion in real value over time: in the time period considered, only Piemonte, Lombardia, Toscana, Molise, Basilicata, Puglia and Sardinia made adjustments to their levels of taxation by raising them. In Piemonte levels of taxation increased considerably from €10.33 per tonne to €25 per tonne. In Sardinia the landfill tax increased from €15.50 to €25.8 per tonne, the highest level in Italy. In Molise tax levels doubled from €10.50 to €21 per tonne. In the remaining regions taxation levels increased only slightly.

There are quite wide differences among regions: the average over the considered period was €14.9 per tonne of MSW landfilled. Piemonte, Basilicata, Veneto, Sardinia and Umbria have the highest levels of taxation at €25 or more per tonne of MSW, while taxes are lowest in Valle d'Aosta and Campania at €5.17 per tonne.

These data were generally confirmed in the telephone interviews. The regions of Campania, Lazio, Molise and Sicily were not straightforward.

Campania is the most densely populated region of Italy, and waste management has always been a problem. Illegal and inappropriate treatment of urban and industrial waste has resulted in contamination of the soil, the atmosphere, and surface and underground water. In February 1993, the first Regional Waste Management Plan was approved to reduce landfill use. However, this plan has not been truly effective and, in 1994, after landfill sites had reached saturation point, a state of emergency was announced.

Table 1.2 Landfill taxes in Italy by region 1999–2008 (€ per tonne)

Region	*Tax range 1999–2008*
Piemonte	10.33–25.00
Valle d'Aosta	5.17
Lombardia	12.91–15.49
Trentino Alto Adige	11.36
Veneto	25.82
Friuli Venezia Giulia	15.49
Liguria	10.33
Emilia Romagna	18.08
Toscana	15.49–16.98
Umbria	25.82
Marche	15.49
Lazio	12.91
Abruzzo	20.52
Molise	10.50–21.00
Campania	5.17
Basilicata	11.00–25.00
Puglia	11.00–15.50
Calabria	10.33
Sicilia	12.36
Sardegna	15.50–25.80

Note: data for 2010–2012 are available upon request.

Our interviewee responsible for waste management in the Lazio region informed us that Lazio had been put under the administration of an external commissioner since 1990. This was due to the not implementation of the conditions of Law 549/95. A succession of administrators focusing on the efficient management of waste has failed to bring an end to the state of emergency. Landfill sites are still saturated and an alternative means of waste disposal is needed urgently.

The history of regulations relating to landfill in Molise is very complex. The first regional law fixing the level of taxation on waste was implemented in 2003, and modified in 2004, 2005 and 2012. In 2005, the level of taxation was raised from €10.50 to €21 per tonne. However, our regional informant told us that a state of emergency had been in place in Molise in relation to waste since 2002, for which reason it had taken eight years after the first national law was passed in 1995 for this region to develop its taxation plan. In particular, the region was unable to formulate a landfill tax plan until central government identified sites for potential landfill. As of August 2012, the situation in this region remains unresolved.

The situation in Sicily is even more complicated, in part because Sicily is an autonomous region that has special status. In 1998 the President of Sicily highlighted the severe crisis in urban waste disposal: he defined the problem as a sanitation and medical emergency, and declared that it was becoming a threat to public order. The root of the problem was the regional plan for waste disposal, which should have been implemented in 1989 but almost ten years later was still embryonic. Also, Sicily's waste disposal and recycling technology was obsolete, inadequate and inefficient. It was common procedure to identify emergency landfill sites in different provinces, and in 1998 there were more than 300 landfill sites in the region. In 1999, Sicily was put under compulsory administration, and its president was appointed the delegated administrator and was mandated with the task of creating and organizing an emergency plan, known locally as PIER.[3] It was supposed to restructure the entire waste system by 2006. However, our interviewee told us that management and organization of the waste sector had been assigned to the regional department for the environment. The situation has improved since 1999: the 300 waste sites have been reduced to 12, and there is coordination between disposal and processing sites.

Furthermore, we note that various increases in the landfill tax rate were observed after 2008 in many Italian regions. Tax levels generally have increased, possibly because of the more stringent targets set by the 2008 Waste Framework Directive and the higher social costs related to landfill. The levels of taxes in Campania might be related to its poor waste management and disposal performances. Many legal and illegal landfill rents remain in place and there are several barriers to a proper landfill diversion process.

Apart from the UK landfill tax, which is correlated to the marginal costs of landfilling and is set according to a time increasing escalator, most environmental taxes are not redefined regularly. Examples include the Danish weight-based packaging tax, which was introduced in 2001 based on life cycle social cost accounting for materials and has never been updated. Revenue generation is often the main motivation for imposing waste taxes. Future research could test the extent to which the 'quality of the tax reform' (e.g. how revenues are recycled – to fund sustainability projects, to reduce labour costs, etc.) affects landfill diversion. To our knowledge, our dataset is unique in providing detailed regional information on landfill taxes in a large country.

Environmental and socio-economic data

Our analysis uses Italian regional waste data that includes observations for all 20 Italian regions in the period 1999–2008. Waste related data are from the Italian environmental agency waste reports (APAT, 2001, 2002, 2003, 2004, 2005, 2006, 2007, 2008); economic data (except landfill taxes) are from the Italian national institute for statistics (ISTAT). Waste management in Italy has changed from landfilling being the main means of disposal

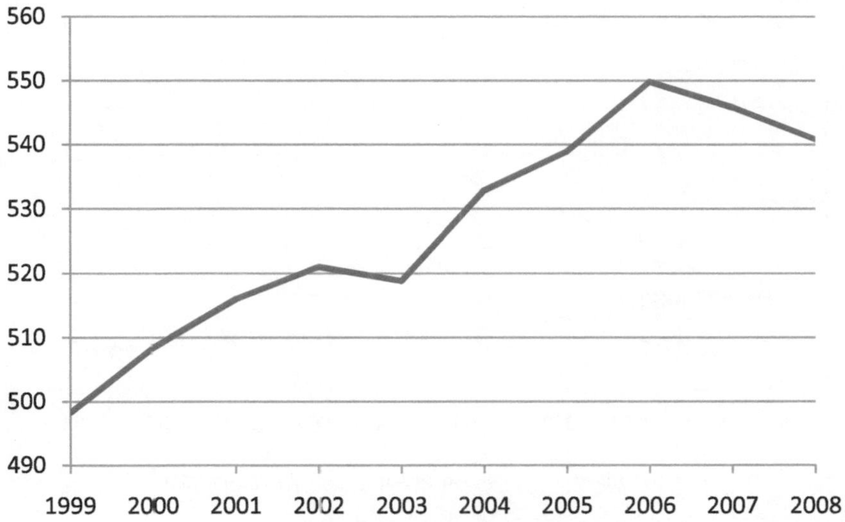

Figure 1.4 Municipal waste generated per inhabitant (kg) in Italy in 1999–2008.
Source: APAT/ISPRA.

to recycling playing an important role. Nicolli (2012) points out that the historical distances between different areas of the country in terms of waste management performance (the north is more advanced and more efficient than the south) are reducing.

Figure 1.4 shows trends in municipal waste generation per inhabitant from 1999 to 2008. We can see that municipal waste production per capita increased constantly between 1999 and 2006, and then shows a slow decrease. However, if we consider the entire period, per capita production of waste grew from 500 kg in 1999, to 540 kg in 2008. However, if we fix the first year as equal to 100, Figure 1.5 shows that, over the analysed period, the amount of waste going to landfill decreased by more than the 25 per cent, from around 380 kg per inhabitant in 1999 to some 260 kg per inhabitant in 2008. At the same time, recycling has increased exponentially, and accounted for some 30 per cent of total waste disposal in 2008 compared to only 13 percent in 1999.[4] Incineration has increased by more than 25 per cent in the period 1999 to 2008, and has assumed an important role in the waste management system.

Figures 1.6, 1.7 and 1.8 depict longitudinal comparisons that highlight the differences among waste management choices across Italian regions in 2008. They reveal a very complex and diverse picture. Figure 1.6 shows that the north of the country is less reliant on landfill than the centre and Sicily, where waste disposal is mostly landfill. Figure 1.7 highlights differences across the north and the south of the country in adoption of separated waste collection. All the regions in the north (excluding Liguria) show amounts of separated collection exceeding 179 kg per inhabitant in

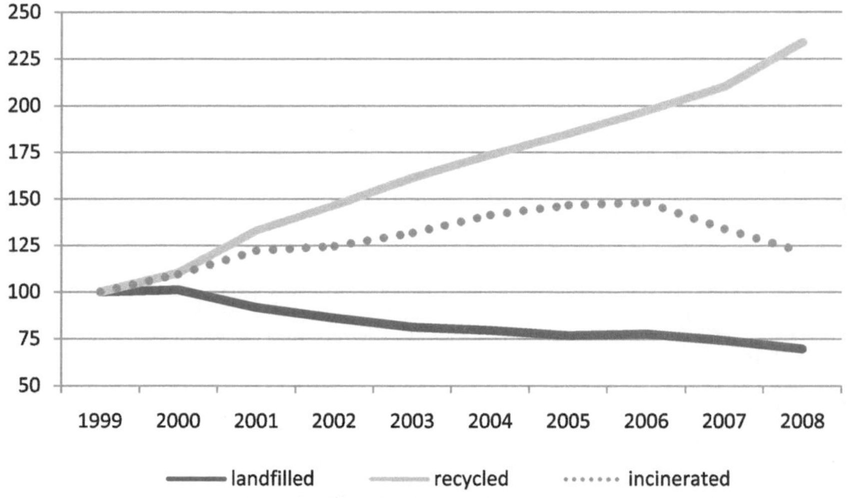

Figure 1.5 Italian waste disposal options trend, year 1999=100.
Source: APAT/ISPRA.

2008, with particularly high levels in Piemonte, Emilia-Romagna, Trentino Alto-Adige and Veneto. In southern Italy there is much less separated collection, with Sicily, Calabria, Puglia, Basilicata and Molise in the lowest tail of the distribution. Figure 1.8 shows incineration levels per province in 2008. In this figure the geographical unit is province rather than region because of the wide differences in incineration adoption. Regional data would have shown a biased picture. Figure 1.8 shows a concentration of incineration plants in the north of Italy, especially Emilia-Romagna and Lombardia. If we exclude these two regions, there is no clear pattern for the use of this technology. The choice to use incineration seems to be a provincial choice rather than a regional strategy.

In line with the literature on Waste Kuznets Curve (WKC) (Mazzanti and Zoboli, 2009) we can formulate the usual general specification:

$$\text{Log (landfilled waste)}_{it} = \alpha_{it} + \beta_1 \text{Log (GDP)}_{it} + \beta_2 \text{Log (landfill tax)}_{it} + \beta_3 \text{Log (Z)}_{it} + \varepsilon_{it}$$

where the first term is an intercept that controls for country fixed effects, the dependent variable is measured as kg of landfilled waste per capita, and the explanatory variables include GDP per capita (β_1), landfill tax (β_2), and a set of variables that control for the regional waste management characteristic based on the information summarized in Table 1.3. Z includes structural factors such as population density. Descriptive statistics and a brief description of the variables are provided in Table 1.3. All the variables are expressed in logarithmic form in the analysis.

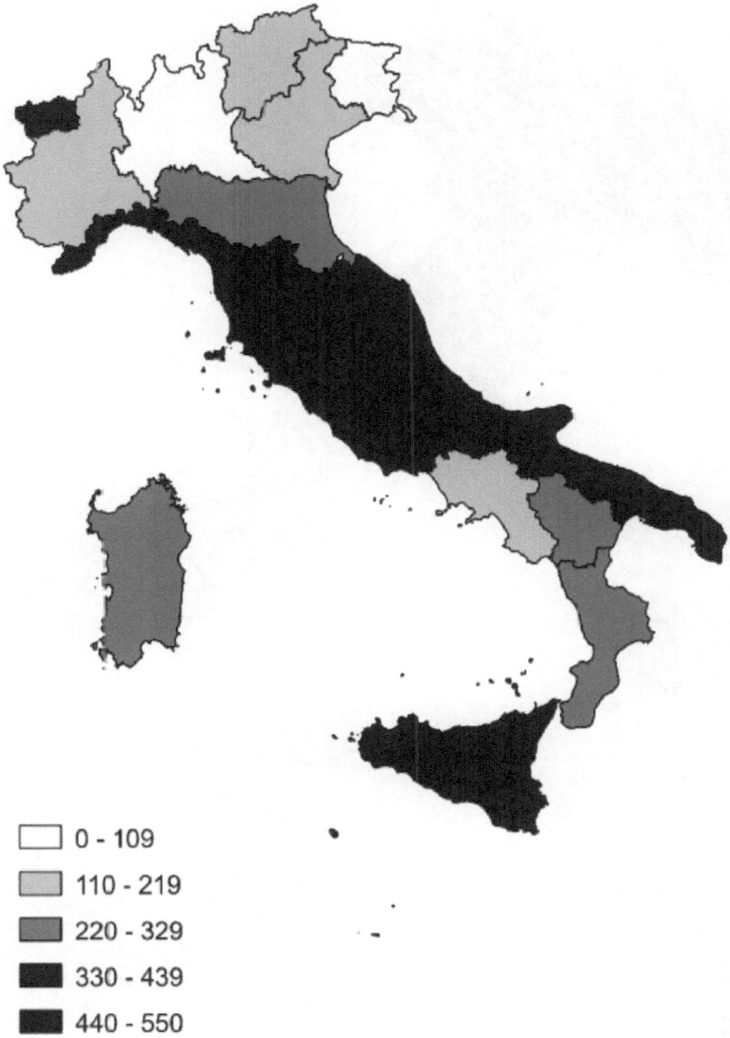

Figure 1.6 Landfilled waste – regional comparison (kg) in 2008.

Econometric evidence

The log-log model results are summarized in Table 1.4. We perform fixed effect (FE) and random effect (RE) estimations as is usual for panel data. If we compare models I (RE) and I (FE), we can see that a Hausman test shows a preference for the FE model, suggesting possible bias in the RE coefficient. Therefore, we present only the FE estimations,[5] which however are similar to the RE ones, especially if we account for waste management characteristics. A first relevant result, which is in line with previous evidence (EEA, 2009),

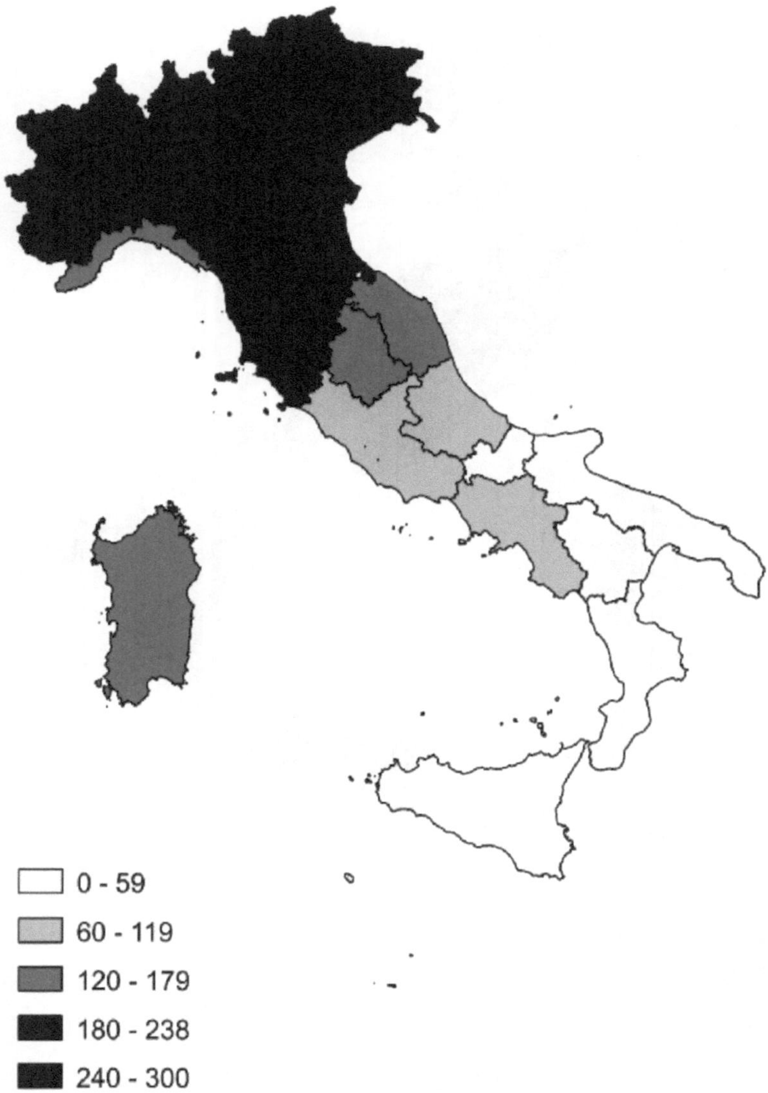

Figure 1.7 Recycled waste – regional comparison (kg) in 2008.

is the non-significance of GDP per capita. This is especially evident if we include other factors in a multivariate analysis. For this reason, we do not include the value added coefficient in the regression table.[6] Also in line with previous evidence is the prominent role of population density in promoting landfill diversion. Economic and health related opportunity costs associated with higher levels of urbanization are confirmed as one of the main drivers of landfill reduction (see similar studies at provincial level Mazzanti *et al.*, 2012). More specifically, a 1 per cent increase in population density leads

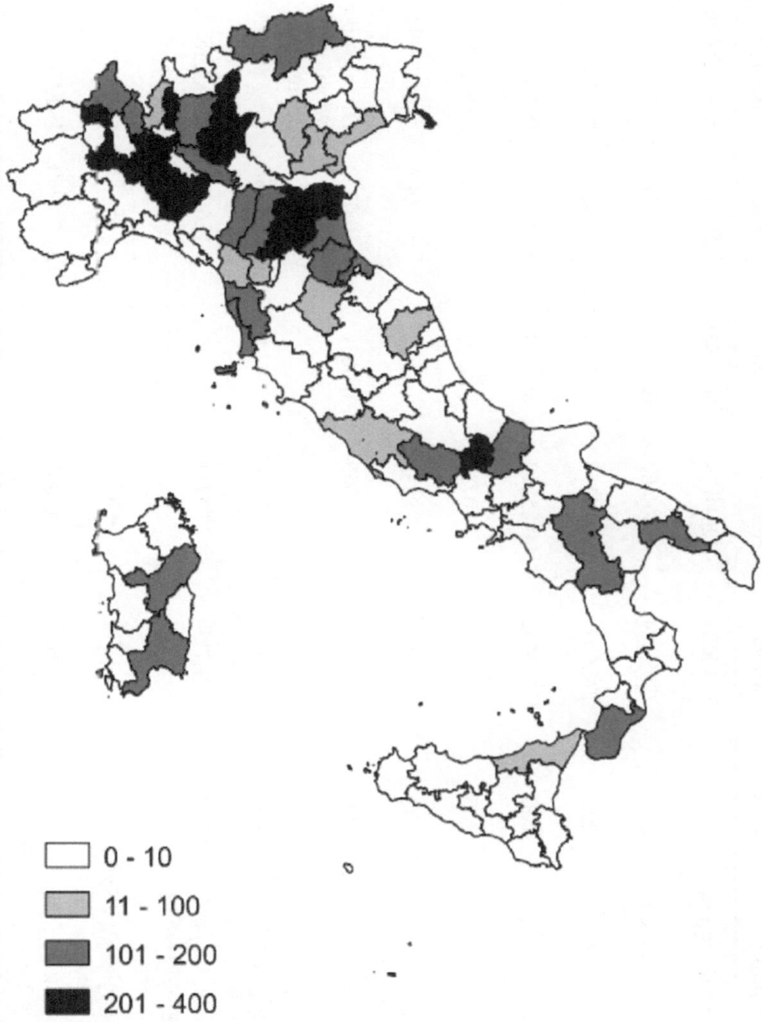

Figure 1.8 Incinerated waste – provincial comparison (kg) in 2008.

to a 4.9–9.1 per cent decrease in the amount of waste that is landfilled. On the other hand, model I (FE) underlines an important difference between our results and previous evidence for provinces. In our analysis, tourist flows are not significant, although previous analyses show that they work to amplify the effect of population density. Although at the provincial level high dependence of the economy on tourism promotes landfill diversion, in the present analysis this effect is less clear. The provincial result might be driven by a few extreme cases, such as Rimini and Venice among others. Our core specification shows that when opportunity costs and potential

Table 1.3 Descriptives and data sources

	Variable description	*Obs*	*Mean*	*St. Dev.*	*Min.*	*Max.*	*Source*
Landfilled	Landfill waste, kg per capita (log in the analysis)	200	344.8817	120.4124	41.9154	618.2991	APAT/ISPRA
GDP	GDP per capita, 1999 thousand euro (log in the analysis)	200	22687.84	5866.67	12423.5	34154.6	ISTAT
Popdens	Population density, inhabitants/km^2 (log in the analysis)	200	177.6093	106.1145	36.43	427.7	ISTAT
Tourism	Total touristic presences (log in the analysis)	200	17525640	14885240	554459	6.15e+07	ISTAT
Incinerated	Incinerated waste, kg per capita (log in the analysis)	200	35.8279	48.0735	0	185.7825	APAT/ISPRA
Recycling	Share of recycling on total waste management (log in the analysis)	200	18.8507	13.8835	0.7	56.8	APAT/ISPRA
Landfill tax	Landfill tax, euro per kg (log in the analysis)	200	0.0149	0.0059	0.0051	0.0258	Direct survey/ websites/official documents
CONTtax	Landfill taxation in contiguous regions (log in the analysis)	200	0.0148	0.0056	0	0.0258	
CONTinc	Average incinerated waste, kg per capita in contiguous regions (log in the analysis)	200	31.1150	32.1241	0	126.258	APAT/ISPRA
Soccap	Electoral turnover share (at provincial level, %) (log in the analysis)	200	82.0263	4.6901	70.085	89.275	Home Ministry
Employment	Employment/inhabitants (log in the analysis)	200	0.40	0.065	0.2716	0.4960	ISTAT

Table 1.4 Estimation results

	I (RE)	*I (FE)*	*II*	*III*	*IV*	*V*	*VI*
GDP	−1.1670***	−0.5177					
Popdens	−0.3698	−4.9285**	−6.4743***	−5.8986***	−8.7874***	−9.1953***	−6.7671**
Tourism	0.1285	0.0012					
Landfill tax			−0.17391**	−0.1754***	−0.20544***	−0.1889***	−0.1717**
CONTtax				−0.2247			
Incinerated					−0.2015***	−0.2232***	−0.2243***
CONTinc						0.1130	
Recycling (t-1)							−0.0077
Region fixed effect	No	Yes	Yes	Yes	Yes	Yes	Yes
Hausman		22.49 (0.0001)					
Instruments							
Overid test							
N	200	200	200	200	200	200	180

Notes: cluster-robust standard error, cluster unit: region. **,*** for p-values <0.05 and <0.01 respectively.

economies of scale are driven by population density, volumes of landfilled waste are lower and probably other forms of disposal are more relevant, while the effects of GDP and tourism are weak and not relevant.

The effect of the regional landfill tax, included in model II below, is negative and highly significant. To our knowledge this is the first study that includes a continuous variable able to account for the stringency of such policy instruments, and its statistical significance in the specifications presented underlines the importance of these instruments for promoting landfill diversion, and more generally in contributing to the transformations in the Italian waste sector since 2002 (see Mazzanti *et al.*, 2012). This result is especially interesting if we consider that the average rate of the Italian landfill tax (€15 per tonne of MSW in 2008) is much lower than in other European countries such as the UK, where it was around £40 per tonne in the same year.

If we look at the characteristics of local waste management two elements stand out. First, the presence in the region of an incineration plant contributes to landfill diversion, indicated by the negative and statistically significant coefficient of the incineration variable in model IV. However, the level of recycling is not a significant driver of landfill diversion (see model VI). Both landfill sites and incineration plants are long-term investment decisions which tie the municipality to the technology for several years, and the choice to build a new incinerator is often at the expense of constructing a new landfill site. On the other hand, more flexible recycling schemes do not seem to have an effect on landfill capacity, and likely are implemented to increase overall disposal capacity, and not to be a direct substitute for total landfilling. On the other side, CONTtax and CONTinc, respectively average level of tax and average level of incineration in contiguous regions, do not seem to influence regional choices related to waste management. These variables are meant to correct for the potential attractiveness of adjacent regions, but the eventual presence of waste trade flows is apparently not influenced by these aspects.

This analysis of 20 Italian regions confirms the relevant role played by landfill tax in the promotion of landfill diversion. We used a unique dataset on the history of landfill taxation and provide the first evidence of its effectiveness using a wide and relevant set of panel data.

Conclusions

In this chapter we analysed the effect of a landfill tax in the context of decentralized regional implementation. To our knowledge, this is the first empirical evaluation of landfill taxes that takes account of cross-section heterogeneity and time dynamics. Our panel data refer to 1999–2008 and are taken from official sources and information derived from interviews with relevant regional officials. Implementation of the data at various levels and across various times confirm our research hypothesis that higher levels and increasing levels of taxation drive down landfilling of municipal waste per capita. Fixed effect models that capture regional idiosyncratic effects show that increasing landfill taxation significantly reduces landfilling. Economic significance and statistical significance are both large. We found also that population density is a 'market' factor that drives down landfilling through the action of land opportunity costs and health costs. Intensity of incineration in the region also drives down landfill. Thus, policy, economic and technological factors all play a role. From a spatial management and policy perspective, the policies of neighbouring regions relating to landfilling and incineration do not influence regional performance. Each region acts independently; there are no interaction effects.

This first evidence of the effectiveness of landfill taxes, which is based on a large relevant dataset, shows that coordinated increases in landfill taxation need to be complemented by stricter waste management strategies if EU targets for management and disposal of municipal waste are to be achieved.

Notes

1 ISTAT is the Italian National Institute of Statistics, while ISPRA/APAT is the Italian Institute for Environmental Protection and Research.
2 The full dataset for 1996–2012 landfill tax rate is available upon request.
3 The PIER is the regional document identifying regional waste management priorities.
4 Decree 22/97 fixed national targets for separated collection; Italy was supposed to achieve 15 per cent of separated collection in total MSW by March 1999, 25 per cent by March 2001, and 35 per cent by March 2003. Decree 152/06 fixed new and more stringent targets (35 per cent of separated collection in total MSW by 31 December 2006, 45 per cent by end December 2008 and 65 per cent by end December 2012). Although the country as a whole has not been able to meet these targets, some regions have performed better than average (see Figures 1.6, 1.7 and 1.8).
5 RE results are available upon request.
6 In line with previous WKC studies, we also tested the squared value of GDP. Since it was generally non-significant, the results are not included here.

References

APAT (2001) *Rapporto Rifiuti 2001 (Waste Report).* Rome: Ministry of the Environment.

APAT (2002) *Rapporto Rifiuti 2002 (Waste Report).* Rome: Ministry of the Environment.

APAT (2003) *Rapporto Rifiuti 2003 (Waste Report).* Rome: Ministry of the Environment.

APAT (2004) *Rapporto Rifiuti 2004 (Waste Report).* Rome: Ministry of the Environment.

APAT (2005) *Rapporto Rifiuti 2005 (Waste Report).* Rome: Ministry of the Environment.

APAT (2006) *Rapporto Rifiuti 2006 (Waste Report).* Rome: Ministry of the Environment.

APAT (2007) *Rapporto Rifiuti 2007 (Waste Report).* Rome: Ministry of the Environment.

APAT (2008) *Rapporto Rifiuti (Waste Report).* Rome: Ministry of the Environment.

Brueckner, J. (2003) 'Strategic interactions among governments: an overview of empirical studies', *International Regional Science Review*, 26: 175–88.

Burnley, S. (2001) 'The impact of the European landfill directive on waste management in the United Kingdom', *Resources, Conservation and Recycling*, 31: 349–58.

Davies, B. and Doble, M. (2004) 'The development and implementation of a landfill tax in the UK', in OECD, *Addressing the Economics of Waste*, Paris: OECD, pp. 63–80.

D'Amato, A., Mazzanti, M. and Nicolli, F. (2011) 'Waste sustainability, environmental management and mafia: analysing geographical and economic dimensions', CEIS Research Paper n. 213, Rome: CEIS – University Tor Vergata.

DEFRA (2005) *A Study to Estimate the Disamenity Costs of Landfill in Great Britain.* London: DEFRA.

EEA (2009) *Diverting Waste from Landfill*, EEA report 7. Copenhagen: European Environment Agency.

Jacobsen, H., Mazzanti, M., Moll, S., Simeone, M.G., Pontoglio, S. and Zoboli, R. (2004) 'Methodology and indicators to measure decoupling, resource efficiency, and waste prevention', ETC/WMF, European Topic Centre on Waste and Material Flows, Copenhagen: European Environment Agency.

IVM (2005) *Effectiveness of Landfill Taxation.* Amsterdam: IVM Institute for Environmental Studies, Vrije Universiteit.

Martin, A. and Scott, I. (2003) 'The effectiveness of the UK landfill tax', *Journal of Environmental Planning and Management*, 46: 673–89.

Mazzanti, M. and Montini, A. (2009) *Waste and Environmental Policy.* London: Routledge.

Mazzanti, M. and Nicolli, F. (2011) 'Waste dynamics, decoupling and ex post policy effectiveness: evidence from the EU', *International Journal of Global Environmental Issues*, 11: 61–78.

Mazzanti, M. and Zoboli, R. (2009) 'Waste generation, incineration and landfill diversion: de-coupling trends, socio-economic drivers and policy effectiveness in the EU', *Environmental & Resource Economics*, 44: 203–30.

Mazzanti, M., Montini, A. and Nicolli, F. (2011) 'Embedding landfill diversion in economic, geographical and policy settings', *Applied Economics*, 43: 3299–311.

Mazzanti, M., Montini, A. and Nicolli, F. (2012) 'Waste delinking, convergence and spatial effects', *Journal of Environmental Planning and Management*, 55: 563–81.

Mazzanti, M., Montini, A. and Zoboli, R. (2008) 'Municipal waste generation and socio economic drivers: evidence from comparing northern and southern Italy', *The Journal of Environment & Development*, 17: 51–69.

Morris, J. and Read, A. (2001) 'The UK landfill tax and the landfill tax credit scheme: operational weaknesses', *Resources, Conservation and Recycling*, 32: 375–87.

Morris, J., Phillips, P. and Read, A. (1998) 'The UK landfill tax: an analysis of its contribution to sustainable waste management', *Resources, Conservation and Recycling*, 23: 259–70.

Nicolli, F. (2012) 'Convergence of waste-related indicators of environmental quality in Italy', *Environmental Economics and Policy Studies*, 14: 383–401. DOI: 10.1007/s10018-012-0042-6.

OECD (2002) *Indicators to Measure Decoupling of Environmental Pressure from Economic Growth*. Paris: OECD.

OECD (2003) *Response Indicators for Waste Prevention within the OECD Area*. Paris: OECD.

Pearce, D.W. (2004) 'Does European Union waste policy pass a cost benefit test?', *World Economics*, 5: 115–37.

Phillips, P., Adams, K., Read, A. and Green, A. (2007) 'Regional variations in waste minimizations in England: challenges and issues for policy development', *Regional Studies*, 34: 297–302.

Stern, D. (2004) 'The rise and fall of the Environmental Kuznets curve', *World Development*, 32: 1419–38.

2 Measuring the impact of economic incentives in waste sorting

Alessandro Bucciol, Natalia Montinari, Marco Piovesan and Lorenzo Valmasoni

Introduction

How to increase waste sorting and achieve more effective waste management are two major items on the policy agenda. At the international level, the importance of regulating waste management is increasing. For instance, in 2006 the European Commission Waste Framework Directive (revised in 2008) stated that EU Member States must recycle at least 50 per cent of their municipal waste by 2020, and in 2005 the US Environmental Protection Agency set a national goal of recycling 35 per cent of the country's municipal solid waste by 2008 (Miller 2006). These actions have been accompanied by citizen uprisings against the construction of new incinerators and landfills and promoting more extensive waste sorting.

Governments, agencies and local municipalities have tried to respond through new collection systems and pricing systems based on monetary and non-monetary incentives. The objective is to increase the cost of 'bad' behaviour (e.g. through pay-as-you-throw – PAYT – pricing schemes), promote 'good' behaviour and reduce costs for citizens (e.g. by collecting sorted waste door-to-door). The implementation of incentive schemes can be costly. For instance, PAYT schemes require households to be provided with bags or tags or fit their containers, or scales to weigh the waste. If the incentive is not properly designed, its benefit may be insufficient to offset the required investment and administrative costs (Kinnaman 2006).

Designing an effective incentive system for the management of municipal waste is not trivial. Which incentives work best? Under which circumstances? Can different incentives operate together? Are there perverse effects over time? To answer these questions case studies have been conducted – around the world – to analyse and *measure* the effect of different incentives on citizens' behaviour (see Kinnaman 2006, for a survey). The resulting evidence is a constructive first step, but there are many other factors that must be considered, analysed and measured. For instance, social awareness (Evison and Read 2001), intrinsic motivation (Halvorsen 2004), social norms (Viscusi *et al.* 2011) and peer pressure (Shaw 2008) can all have an impact on the effectiveness of a waste management system.

An additional level of complexity is represented by the huge variety of practices observed in the real world. Over the last few decades, most municipalities in the western world have examined and reorganized their waste management systems, combining the basic ingredients, such as pricing and collection systems, and introducing small changes to personalize resulting waste management systems. It is difficult to categorize these ad hoc waste management systems and similarly difficult to compare their results.

For instance, with respect to waste collection systems, there is a continuum of practices between two extreme solutions (drop-off versus door-to-door – DtD), that includes the different types (and volumes) of sorted waste fractions that will be collected; how to define recyclable materials; frequency of their collection; and types and sizes of containers. The range is huge. Similarly, pricing systems come in many different variants – as discussed in the next section. Thus, comparing different schemes and assessing the factors that contribute to or reduce their effectiveness, is very difficult, although some sort of evaluation is necessary in order to identify best practice and produce good guidelines.

This chapter contributes to the debate on ways of collecting and analysing the data on waste sorting. We argue that an experimental methodology or controlled analysis is appropriate. To assess the effectiveness of policy instruments for promoting certain behaviour, we need to compare the group at whom the policy intervention is directed with a similar group with no policy intervention. This type of analysis can be described as a controlled analysis because it allows the change in behaviour induced only by the policy intervention to be isolated from other factors.

In the next section, we present evidence of best practice in municipal waste management, from Europe and the United States, focusing on door-to-door collection combined with a PAYT incentive system. Then we discuss why controlled analysis should be used to estimate the effect of different policies. Afterwards, the section devoted to the Treviso district case draws on Bucciol *et al.* (2011) to show why controlled analysis represents progress towards a better understanding of the effects of different policies, as opposed to the biased results from uncontrolled analysis.

Incentive schemes in waste management

Economic incentives are becoming more widespread in the context of waste management in order to decrease the amount of domestic waste produced and to increase the amount sorted by households. Economic incentives may be monetary or non-monetary. An example of a non-monetary incentive is *kerbside collection*; many municipalities are abandoning drop-off collection systems and replacing them with DtD collection to encourage households to sort their waste. DtD encourages household sorting by reducing the cost to the citizen (Jenkins *et al.* 2003). This method became popular in the United States in the mid-1980s and in 1988 there were about 1,000 kerbside collection programmes in the United States, and around 9,000 in 1996. Jenkins *et al.* (2003), using US national household level data, found that kerbside programmes had a positive impact on household sorting behaviour, especially for materials that are expensive to transport to drop-off centres (e.g. dense and bulky materials). For Europe, Cotterill *et al.* (2008) found that in Manchester, UK, the implementation of a DtD programme increased household sorting by about 7–8 per cent.

Debate continues on the effectiveness and overall economic convenience of monetary incentives – including subsidies paid to producers who use recycled material, and taxes on virgin materials, which can increase producers' demand for recycled inputs, raising the price for recycled materials and resulting in economic benefits for households that sort their waste. Deposit-refund systems are another way of inducing households and producers to take account of the external costs associated with waste disposal. However, the most pervasive monetary incentive is a direct fee based on waste volumes, that is, a PAYT or unit-based pricing scheme, which forces households to carry the full social costs of their disposal decisions, inducing more efficient choices.

PAYT is usually implemented in weight-based, volume-based or subscriptions programmes. In weight-based schemes, the collector bills the household according to the weight of its waste. Volume-based schemes usually require households to purchase waste bags or stickers/tags that they can attach to their waste containers, alternatively, they may be billed according to the number of their bins. Subscription schemes require the householder to (pre-)pay for a predetermined number of bins or a container of a predetermined size.

The first PAYT schemes were implemented in San Francisco in 1932, and in Austria in 1945 (Reichenbach 2008). However, in both Europe and the United States it was not until the mid-1960s that such schemes became more widespread, when the technology for waste collection services, such as transponder technology for electronic identification, allowed municipalities to charge for waste on a variable base.

Since the early 1990s, more and more municipalities in North America have introduced PAYT schemes. In the late 1980s there were about 100 in operation, and this had risen to 1,000 in 1993 and 5,200 in 2001. In 2006 PAYT programmes had been implemented in almost 7,100 jurisdictions in the United States, covering some 25 per cent of the US population. Minnesota, Iowa, Wisconsin, California, New York, Washington and Pennsylvania are the states with the highest number of PAYT schemes in operation – more than 200 in some cases. The effects of PAYT schemes in the United States have been examined in a series of studies based on data from more than 1,000 municipalities (Skumatz and Freeman 2006). Hong *et al.* (1993) used survey data on some 2,300 households in Portland, United States, on waste sorting activity, and costs of waste collection. They found that a volume unit-based user fee increased household sorting, but found no evidence of an effect on the amount of waste produced. Fullerton and Kinnaman (1996) used data on 75 households measured before and after the implementation of a volume-based scheme in Charlottesville, United States, showing that, after introduction of a fee, the amount of sorted waste increased, the overall weight of waste produced decreased slightly and the volume of waste (measured as number of bags or cans) decreased quite substantially.

In Europe, PAYT schemes are very widespread and in Switzerland and Luxembourg, for example, they are countrywide (Reichenbach 2008). PAYT was boosted by European legislation and the 2008/98/EC directive on prevention and recycling as immediate objectives. In most countries, legislation based on the polluter-pays principle,[1] has induced many municipalities to implement PAYT systems and there are many successful examples of this pricing system in Europe. For instance, in 1999 in Landkreis Schweinfurt, Germany, a multi-tier pricing scheme was introduced which is a combination of the weight and volume pricing systems. It involves an annual fixed fee and a fee per bin (of whatever size) emptied, and a two-tier weight-based fee, higher for residual waste (that cannot be composted, recycled or reused). The introduction of this scheme entailed a total reduction in waste of around 13 per cent, a reduction in residual waste of about 46 per cent, and the average sorted/total waste rate increased from 64 per cent to 76 per cent. A PAYT scheme was introduced in 1998 in Ghent, Belgium. The scheme varies across the city. Urban centre residents have to purchase waste bags from local stores. Residents in rural areas use bins and pay in advance for a certain number of collections. It resulted in increased participation among households, increased sorted waste collection, and a slight slowing of the overall growth in waste production (Hogg 2006). Allers and Hoeben (2010) conducted a study based on 458 Dutch municipalities and found evidence that the price effect depends on the pricing system: weight-based systems were more effective in reducing the amount of waste than volume-based systems. User fees increased sorting, but not to the same extent as they reduced amounts of waste.

Asia has a number of PAYT systems in operation. South Korea was the first country to introduce a countrywide PAYT scheme, in 1995 (Hogg 2006). It adopted a bag pricing system, allowing municipalities to charge different fees. The amount of residual waste decreased significantly and sorted waste increased twofold between 1994 and 2004. However, the scheme also led to illegal dumping and disposal of waste. In Japan PAYT programmes first became popular in the early 1970s and had a revival in the 1990s. Mainly municipalities with fewer than 30,000 inhabitants adopted PAYT, and achieved total reductions in waste of between 20 per cent and 30 per cent (Sakaia *et al.* 2008).

From these few examples, we can see that there are many different variations within the PAYT programs, but that all these schemes were aimed at increasing waste sorting by the householder. Some municipalities impose a quantity-based fee, others apply fees with fixed and variable components. Also, PAYT schemes can be implemented in combination with DtD collection systems. It is difficult, therefore, to evaluate the different effectiveness of collection and pricing systems, since increased sorting may stem from several factors, such as the structure of the fee and parallel pricing and collection systems.

Different populations may have certain characteristics that induce sorting behaviour regardless of an incentive programme. For instance, in high population density municipalities, households may face high storage costs because of their more restricted living space. This may result in higher levels of waste, with or without a PAYT and/or DtD system. Household income level also has an effect. Low-income households, such as unemployed and immigrants, may always produce less waste regardless of any specific programmes.

Environmental awareness is also important. Several programmes have been introduced in response to awareness of environmental issues. Sorting tends voluntarily to increase when DtD programmes are implemented. A combination of different approaches and the characteristics of the population make it difficult to compare waste management systems and to understand the effectiveness of sorting incentives.

In the following two sections we discuss the importance of assessing the impact of a policy intervention and review the main issues related to this assessment. We discuss the case of the Treviso district in Italy as an example of a natural experiment for evaluating the impact of monetary and non-monetary incentives on waste sorting.

How to evaluate different policies?

The essential ingredient in any impact evaluation is the availability of data that can be used to measure changes in targeted behaviour and capture the effects of factors that might be influencing this behaviour. In the absence of data, the boldest theory is only an untested hypothesis. The more reliable and rich the collected data, the better will be the possibility afforded to test whether a particular policy is having the desired results, and to isolate its effect on behaviour from the effects of other factors. The availability and quality of data has improved in recent years. This is allowing researchers to measure phenomena, evaluate the impact of alternative policy interventions by isolating them from other confounding factors, and to make more precise predictions.

Evaluating the effectiveness of policy interventions implies assessing the causality of policy on desired behaviour. This essentially requires the ability to measure changes in the behaviour of the targeted group, but we need also to know whether the observed change is due exclusively to policy intervention and is not influenced by other factors. Establishing the causal effect of policy intervention means identifying behaviour changes that would not have occurred had the policy not been adopted. This requires the ability to estimate the net effect of the changes in behaviour that are the effect only of the policy intervention.

For example, to evaluate the effectiveness of a PAYT scheme (or any other type of intervention) in the presence of an increase in the sorting behaviour, we need to isolate the effect of PAYT from the effect of DtD or citizens' environmental awareness, as well as from all other factors that can potentially affect sorting behaviour even in absence of the PAYT scheme.

The question to be addressed, therefore, is what would be happened if the specific policy instrument had not been implemented. This requires a 'counterfactual' scenario that allows us to compare what actually happened and what would have happened without the policy being in place. We can measure the impact of the specific policy intervention by comparing counterfactual outcomes with the outcomes obtained as a result of the intervention.

Let us assume that we want to evaluate the impact of a policy on a targeted behaviour, over two periods of time. Assume also that we have data from two groups that are identical apart from the influence of the policy intervention. In the *treated* group, the policy is implemented (continuous line in Figure 2.1); in the *non-treated* group there is no policy intervention (dashed line in Figure 2.1).

Figure 2.1 shows three graphs depicting three possible situations. In the first graph (a), the treated group passes from 15 per cent at time 1 to 35 per cent at time 2. The difference between the observed change in behaviour at time 2 and time 1 represents the *total* change in behaviour. To understand the net change (i.e. the impact of) due to the policy intervention we need to consider the non-treated group (dashed line). It starts at 15 per cent at time 1 and it goes to 25 per cent at time 2. The change in the dashed line gives us information about the effect of other factors than the policy. Therefore, to calculate the impact of the policy we need to subtract the change in the non-treated group (i.e. 10 per cent) from the total change observed in the treated group (i.e. 20 per cent). From this example we can conclude that the change due to the policy intervention is 10 per cent.

Next we consider the graph in panel (b). In comparing the two groups, it can be seen that the two lines coincide, meaning that the policy intervention has no effect and the observed total change in behaviour must be down to the effects of other factors. In graph (c) both lines show an increasing trend. The continuous line (treated group) starts at 15 per cent at time 1 and goes to 35 per cent at time 2. The dashed line starts from 5 per cent in time 1 and it reaches 25 per cent in time 2. In both groups we observe a 20 per cent increase in the changed behaviour. The net effect of the policy intervention, however, is null, as in the previous case.

This example shows that how the non-treated group is identified and, similarly, how the treated group is composed compared to the non-treated group, are crucial issues. When the two groups are identified based on external factors, such as public policy or other events not related to the evaluation, we refer to a 'natural' experiment and to the non-treated group as the *control group*. In some circumstances, natural experiments are designed by social scientists in order to evaluate the effectiveness of policy intervention

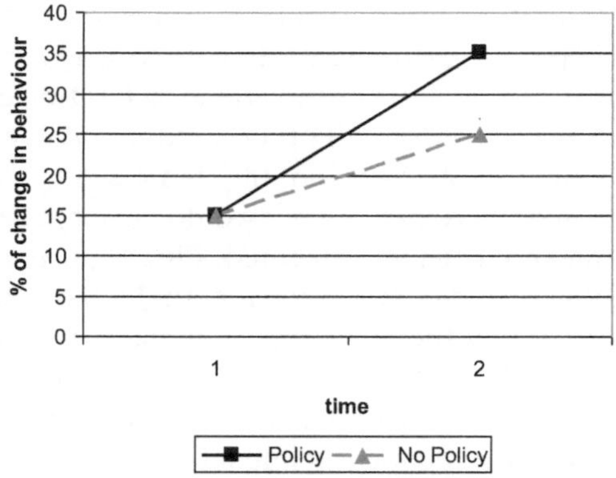

(a) Total change: 20%; net change: 10%. Same starting level.

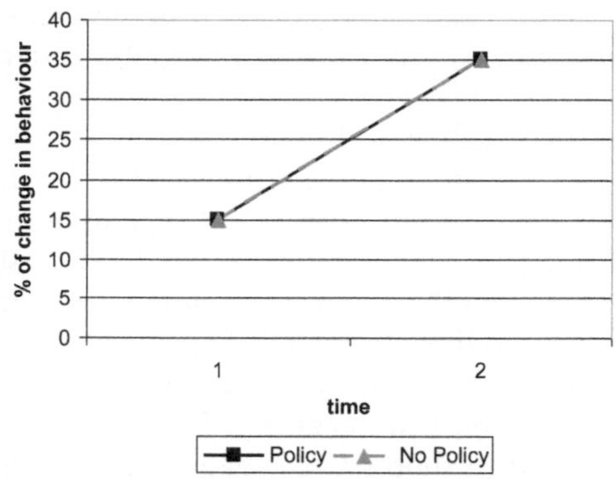

(b) Total change: 20%; net change: 0%. Same starting level.

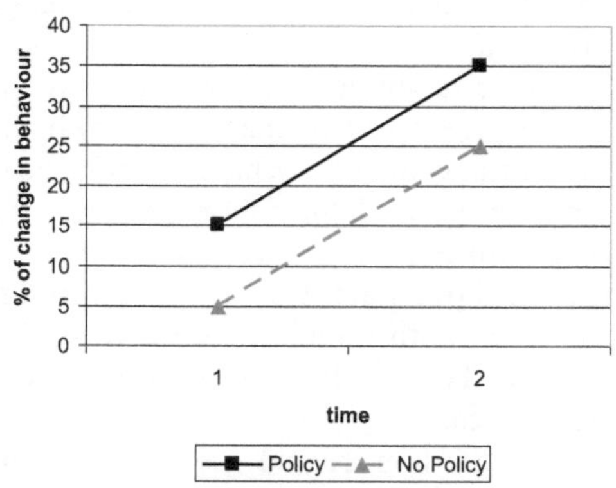

(c) Total change: 20%; net change: 0%. Different starting level.

Figure 2.1 Example of impact evaluations.

by applying experimental methods to the real world. However, natural experiments are not easily found: in particular, it is unlikely that policy changes will be truly random, and, in some cases, randomization may not be possible for ethical reasons. In all cases where counterfactual outcomes cannot be observed directly, the impact analysis is run with *comparison group* to proxy for a control group. The difference between a control group (natural experiment) and a comparison group (*quasi*-natural experiment) relies on the random assignment of observations to the two groups. The main advantage of natural and quasi-natural experiments is that they can be conducted on large samples and in an environment that is natural rather than the artificial environment of many laboratory experiments.

In the next section we analyse the effectiveness of two different waste management systems designed to foster municipal waste sorting: PAYT (monetary incentive) and DtD (non-monetary incentive). Despite the many studies that assess the joint effect of PAYT and DtD (see Kinnaman 2006), to our knowledge, none are based on scenarios that resemble a natural field experiment. Our data are drawn from Bucciol *et al.* (2011), and allow us to disentangle the net effect of PAYT from the effect of DtD, and to assess the measurement bias that standard analysis can produce.

The case of the Treviso district

We describe a natural field experiment exploring the impact of different waste management policies; for further details see Bucciol *et al.* (2011). This case includes the 95 municipalities in the district of Treviso, Italy. Treviso is a relatively wealthy area – per capita GDP of around €30,000 in 2008. It has a population of around 900,000, and a relatively high population density (355 persons per km^2; ISTAT 2009). Its municipalities, which are mostly small in size (70 out of 95 have fewer than 10,000 inhabitants), are close to each other.

Our dataset contains annual information on waste disposal for the years 1999–2008, for each of the 95 municipalities. The municipalities in the sample adopted two different types of waste management policies: collection (drop-off or DtD), and payment system (flat fee or PAYT). Since DtD and PAYT both provide an incentive to sort more, we can group our observations into three categories: no incentive (drop-off and flat fee), single incentive (DtD and flat fee) and double incentive (DtD and PAYT). We know the demographic structure of each municipality and its proximity to other municipalities. These data can be employed for several purposes, and in our case are used to study the net effect of DtD and PAYT on the Sorted Waste Ratio (SWR), that is, the ratio between sorted waste and total (sorted plus unsorted) waste produced.

The dataset is particularly suitable for the present analysis for several reasons. First, it includes a number of different policies along the two dimensions of interest (collection method and payment system). Second, the panel structure of the data allows us to control for time-specific environmental concern (likely to be increasing over the years due to awareness campaigns), and to isolate the potential 'learning effects' from incentives offered over the years. Third, and most importantly, the municipalities in the sample are not directly responsible for the decisions over waste management; in some senses, they can be considered to be participating in an experiment where the environment is set exogenously. We explain this below.

Since the late 1980s municipalities in Italy have been able to implement their preferred waste management policies. In order to benefit from scale economies, the Treviso district chose to delegate policy decisions to consortia of municipalities. Municipalities implemented the decisions of the consortium managing their particular geographical area at different moments in time because some had to delay implementation until a previous contract with an external company came to an end. Once having subscribed to the relevant consortium, the municipality can no longer decide on its waste management policy, but is required to follow the consortium's recommendations, which are formulated by an independent board of managers. Over time, three consortia (Priula, Savno and TV3) have been set up, all similar in size. From a situation where all Treviso municipalities had adopted drop-off collection and flat fee pricing, all three consortia embarked on a gradual introduction of DtD, but with different payment systems. Priula and Savno began implementation of PAYT in 2002, while TV3 continued with flat fee pricing.

At the beginning of the sample period, in 1999, all municipalities had adopted drop-off collection and flat fee pricing and their SWR was around 35.4 per cent, compared to higher percentages in other areas in north Italy (ISPRA 2011). At the end of the sample period, in 2008, 41 municipalities had adopted DtD collection and flat fee pricing and were achieving 66.7 per cent SWR, and 53 municipalities had adopted DtD collection and PAYT pricing and were achieving 74.95 per cent SWR. This result was exceptional compared to the national average of 30.6 per cent (Arpav 2009). Priula is now considered an example of best European practice, and is frequently ranked first in national statistics on waste (Legambiente 2009).

The evidence seems clear. However, it ignores intrinsic differences across municipalities and over time. It also does not take account of the fact that users may be emulating what they observe in adjacent municipalities, or may be becoming 'waste tourists', an extremely negative side effect of PAYT – especially in areas like Treviso where municipalities are very close to one another, and where people frequently commute between municipalities for their work. The objective of the econometric analysis, inspired by Bucciol *et al.* (2011), is to remove these potentially disturbing effects and provide a cleaner measure of the net effect of DtD and PAYT policies.

Table 2.1 reports the main findings from a fixed-effects panel ordinary least squares (OLS) regression analysis with municipality-clustered standard errors, on our data. The purpose is to quantify the change in municipality

Table 2.1 Regression output

	(1)	*(2)*	*(3)*	*(4)*
Dependent variable	*SWR*	*SWR*	*Log(PCSW)*	*Log(PCTW)*
Analysis	*Controlled*	*Uncontrolled*	*Controlled*	*Controlled*
PAYT	6.345***	8.279*	0.236***	0.028
	(1.929)	(4.025)	(0.050)	(0.033)
Years of PAYT	1.387***	–0.224	0.026**	–0.018***
	(0.377)	(1.827)	(0.010)	(0.007)
DtD	9.424***	10.219**	0.082**	–0.099***
	(1.926)	(4.003)	(0.039)	(0.021)
Years of DtD	0.186		0.011	–0.025**
	(0.758)		(0.017)	(0.010)
Adjacent municipalities with double incentive	2.602* (1.382)	–1.935 (4.110)	0.132*** (0.040)	0.019 (0.022)
Adjacent municipalities with no incentive	–4.736*** (1.680)	–14.032*** (3.080)	–0.096** (0.043)	0.062*** (0.021)
PAYT × adjacent municipalities with no incentive	5.018* (2.610)	16.386*** (3.295)	0.052 (0.068)	–0.115*** (0.039)
Log (density)	31.639	82.451	0.744	-0.620*
	(20.387)	(49.587)	(0.622)	(0.356)
% children residents	0.658*** (0.209)	0.123 (0.503)	0.021*** (0.007)	–0.007* (0.003)
% retired residents	0.249 (0.808)	–0.144 (2.020)	0.015 (0.034)	–0.014 (0.016)
% non-native residents	–0.044 (0.382)	0.910 (1.246)	–0.016 (0.016)	0.001 (0.007)
% unemployed residents	0.155 (0.651)	–1.297 (0.836)	–0.069*** (0.023)	–0.013 (0.010)
Time	2.353***	1.803	0.160***	0.056***
	(0.803)	(1.464)	(0.029)	(0.012)
Time^2	–0.197***	–0.169	–0.014***	–0.000
	(0.071)	(0.112)	(0.003)	(0.001)
Constant	–146.617	–448.829	0.128	9.672***
	(120.434)	(312.646)	(3.843)	(2.223)
Observations	950	960	1,660	1,660
R-squared	0.792	0.923	0.782	0.246
Fraction of variance due to municip. effects	0.881	0.973	0.818	0.963

Table 2.1 (cont.)

	(1)	*(2)*	*(3)*	*(4)*
Dependent variable	*SWR*	*SWR*	*Log(PCSW)*	*Log(PCTW)*
Analysis	*Controlled*	*Uncontrolled*	*Controlled*	*Controlled*
R^2	0.257	0.255	0.782	0.246
R^2 within-group	0.792	0.923	0.413	0.230

Notes: SWR: Sorted Waste Ratio; PCSW: Per Capita Sorted Waste (in kg); PCTW: Per Capita Total Waste (in kg). 'Controlled' analysis: sample with the control group. 'Uncontrolled' analysis: sub-sample without the control group, duplicated six times to have comparable size.

Robust standard errors in parentheses. *** p-values<0.01, ** p-values<0.05, * p-values<0.1.

SWR (Columns (1)–(2)) and amount of waste produced (Columns (3)–(4)) as a result of the waste management policies adopted, controlling for time, municipality demographic characteristics, and the policies adopted by adjacent municipalities. The last group of variables includes whether the municipality, in the year of observation, was adjacent to a municipality implementing DtD and PAYT (double incentive), or to a municipality implementing drop-off and flat fee (no incentive). This should capture 'emulation' of peer behaviour. The interaction between PAYT and emulation of peer behaviour is informative about whether PAYT induces some citizens to 'hide' their waste in nearby municipalities with drop-off waste collection, in order to avoid the per-unit fees. Our results for SWR (Column (1)) can be summarized as follows:

- *Effectiveness of incentives.* On average the SWR increases by 9.4 per cent with DtD and by 6.3 per cent with PAYT. Both incentives are effective, although the DtD effect is stronger.
- *Understanding PAYT takes time.* The SWR increased by a further 1.4 per cent for every additional year of PAYT. People need time to fully understand and exploit PAYT. In contrast, there is no time effect for DtD.
- *Increasing awareness of the problem.* In the first six years of the sample, SWR increased over time, regardless of which waste management policy was adopted ($2.353/(2 \times 0.197) = 597$).
- *Peer emulation.* SWR falls by 4.7 per cent if at least one adjacent municipality adopts no incentive; the size of this effect is so large that potentially it could offset the positive effect of PAYT. In contrast, there is no emulation of adjacent municipalities with the double incentive.

Column (2) reports the results of an 'uncontrolled' experiment (i.e. with no control group), where we concentrate only on the subset of 16 municipalities that moved to DtD in 2001 and to PAYT in 2002. The purpose is

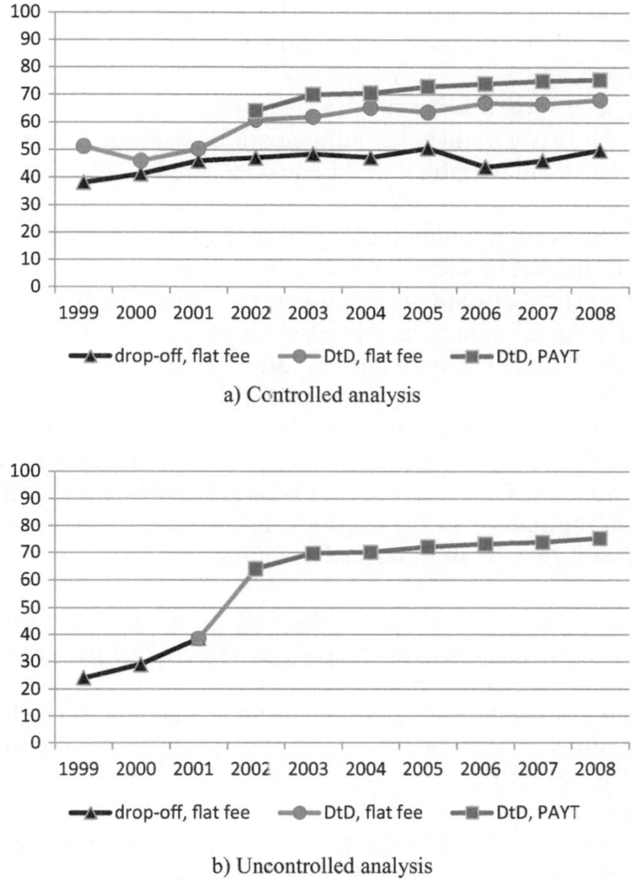

Figure 2.2 Trend in SWR.

to understand what the analysis would suggest in the absence of a control group, which is, the group of municipalities that adopted different policies in the same period. Figure 2.2 shows that, while in the full sample we observe three different waste policies simultaneously (left panel), in this situation we observe the different policies in sequence (right panel), and observe a marked increase in SWR with the change from one policy to another.

In order for the sample size to be comparable to the previous analysis, we duplicate the observations six times, to give a total of 160 × 6 = 960 observations. This analysis has several shortcomings:

- *Biased incentive effect.* We estimate larger effects for both DtD and PAYT, the latter is only weakly significant.
- *Difficult to capture time-related effects.* We no longer find the effects of the variables related to time to be significant. Also, because of the way

the data were collected, we cannot estimate the effect of years of experience of DtD because they coincide with years of experience with PAYT *plus* the PAYT variable.

- *Confounding effects*. We observe suspiciously large effects, with opposite signs, for the two variables for municipalities without incentives. This suggests that these variables are highly correlated.

The uncontrolled analysis highlights the importance of a control group in order to evaluate the impact of a specific policy instrument, such as PAYT.

The SWR is the ratio between sorted waste and total waste. It may increase because: (1) the amount of sorted waste increase, but total waste remains unchanged; (2) there is a reduction in total waste, but sorted waste remains unchanged; (3) there is an increase in sorted waste and a reduction in total waste. We next examine in more depth the composition of the SWR. In Columns (3) and (4) the dependent variables are respectively the logarithm of the amount of Per Capita Sorted Waste (PCSW) in kilograms, and the logarithm of the amount of Per Capita Total Waste (PCTW) in kilograms. The main results can be summarized as follows:

- *PCSW increases with DtD and PAYT.* We find an 8.2 per cent effect of DtD (sorting is simpler) and an even bigger effect, 23.6 per cent, of PAYT (the alternative is costlier).
- *PCTW falls with DtD.* We find a 9.9 per cent effect of DtD (which imposes constraints on the volume of waste and frequency of collection), but no effect of PAYT.
- *Waste production increases over time.* The production of PCSW increased during the first six years ($0.16/(2 \times 0.014) = 5.71$) of our sample, while the production of PCTW kept on growing annually at a linear rate of 5.6 per cent. Environmental economists attribute this to increased wellbeing (Mazzanti and Zoboli 2008).
- *Peer emulation.* Adjacent municipalities without incentives reduces PCSW by 9.6 per cent and increases PCTW by 6.2 per cent; adjacent municipalities with double incentives increases PCSW by 13.2 per cent.
- *Waste tourism.* The amount of PCTW falls by 11.5 per cent when the municipality adopts PAYT and there are adjacent municipalities without incentives.

The analysis contained in Bucciol *et al.* (forthcoming) reports similar results by focusing only on the municipalities belonging to consortium Priula and TV3 for the years 1999–2010.

Conclusions

This chapter contributes to the literature in two ways. First, from a methodological point of view we discuss how data can be collected and analysed to measure the net effects of alternative policy systems, distinct from other factors such as learning or civic awareness, on recycling. Second, by discussing the results based on the data in Bucciol *et al.* (2011), we show the importance of natural field experiments to properly measure the effect of different policies. We find that DtD and PAYT both increase waste sorting, although PAYT needs more time to be exploited in the right way. There is also evidence of large interactions with adjacent municipalities (through emulation of peer behaviour, or waste tourism), which points to the benefits of policies being decided by consortia of municipalities.

This analysis should be informative for policy makers because it should allow them to predict the effect on the population of introducing and maintaining a specific waste management system. Being able to consider the impact of a waste management system on day-to-day life is important and allows progress beyond facts and correlations towards a deeper assessment of the causality between policy and human behaviour.

Notes

1 Namely, 'the cost of measures to reduce pollution according to the extent of either the damage done to society or the exceeding of an acceptable level (standard) of pollution' (OECD, 2001).

References

Allers, M.A. and Hoeben, C. (2010) 'Effects of Unit-based Garbage Pricing: A Differences-in-Differences Approach', *Environmental and Resource Economics*, 45: 405–28.

Arpav (2009) 'Produzione e gestione dei rifiuti nel Veneto – anni 2007–2008', Dipartimento Provinciale di Treviso – Servizio Osservatorio Suoli e Rifiuti. Online. Available at: www.arpa.veneto.it/pubblicazioni/htm/scheda_pub.asp?id=215 (accessed 10 Febuary 2012).

Bucciol, A., Montinari, N. and Piovesan, M. (2011) 'Do Not Trash the Incentive! Monetary Incentives and Waste Sorting', *Harvard Business School Working Paper, No. 11–093*, March, Cambridge, MA.

Bucciol, A., Montinari, N. and Piovesan, M. (forthcoming), 'Do Not Miss the Opportunity! When to Introduce Monetary Incentives' in T. Kinnaman, and K. Takeuchi (eds.), *Handbook of Waste Management*, Cheltenham: Edward Elgar Press.

Cotterill, S., John, P. and Liu, H. (2008) 'How to Get Those Recycling Boxes Out: A Randomised Controlled Trial of a Door-to-door Recycling Campaign'. Online. Available at: www.york.ac.uk/iee/events/Presentations%202008/Sarah%20Cotterill%20%20How%20to%20get%20those%20recycling%20boxes%20out_York_Final.pdf (accessed 10 February 2012).

Evison, T. and Read, A.D. (2001) 'Local Authority Recycling and Waste Awareness Publicity/Promotion', *Resources, Conservation & Recycling*, 32: 275–91.

Fullerton, D. and Kinnaman, T.C. (1996) 'Household Responses to Pricing Garbage by the Bag', *American Economic Review*, 86: 971–84.

Halvorsen, B. (2004) 'Effects of Norms, Warm-Glow and Time Use on Household Recycling', *Discussion Papers 389*, Oslo: Statistics Norway.

Hogg, D. (2006) 'Working Group on Waste Prevention and Recycling. Impacts of Unit-based Waste Collection Charges'. Online. Available at: www.oecd.org/officialdocuments/publicdisplaydocumentpdf/?cote=ENV/EPOC/WGWPR(2005)10/FINAL&docLanguage=En (accessed 10 February 2012).

Hong, S., Adams, R.M. and Love, H.A. (1993) 'An Economic Analysis of Household Recycling of Solid Wastes: The Case of Portland, Oregon', *Journal of Environmental Economics and Management*, 25: 136–46.

ISPRA (2011) 'Annuario sui rifiuti'. Online. Available at: http://annuario.apat.it/capitoli/Ver_0/versione_integrale/Rifiuti.pdf (accessed 19 April 2012).

ISTAT (2009) 'Rapporto Annuale 2008'. Online. Available at: www3.istat.it/dati/catalogo/20090526_00 (accessed 19 April 2012).

Jenkins, R.R., Martinez, S., Palmer, K. and Podolsky, M.J. (2003) 'The Determinants of Household Recycling: A Material Specific Analysis of Unit Pricing and Recycling Program Attributes', *Journal of Environmental Economics and Management*, 45(2): 294–318.

Kinnaman, T.C. (2006) 'Policy Watch Examining the Justification for Residential Recycling', *Journal of Economic Perspectives*, 20: 219–32.

Legambiente (2009) 'Comuni Ricicloni 2008'. Online. Available at: http://risorse.legambiente.it/docs/DossierCR2008.0000001126.pdf (accessed 19 April 2012).

Mazzanti, M. and Zoboli, R. (2008) 'Waste Generation, Incineration and Landfill Diversion. De-Coupling Trends, Socio-Economic Drivers and Policy Effectiveness in the EU', *Working Paper 94.*2008, Fondazione ENI Enrico Mattei, Milan.

Miller, C. (2006) 'States Lead the Way: Pioneering Recycling Efforts in the US', *Waste Management World*. Online. Available at: www.waste-management-world.com (accessed 14 February 2012).

Organisation for Economic Co-operation and Development (OECD) (2001) 'Glossary of Statistical Terms'. Online. Available at: http://stats.oecd.org/glossary/detail.asp?ID=2074 (accessed 19 April 2012).

Reichenbach, J. (2008)'Status and Prospects of Pay-As-You-Throw in Europe: A Review of Pilot Research and Implementation Studies', *Waste Management*, 28: 2809–14.

Sakaia, S., Ikematsub, T., Hiraia, Y. and Yoshidac H. (2008) 'Unit-Charging Programs for Municipal Solid Waste in Japan', *Waste Management*, 28: 2815–25.

Shaw, P.J. (2008) 'Nearest Neighbour Effects in Kerbside Household Waste Recycling', *Resources, Conservation and Recycling*, 52: 775–84.

Skumatz, L.S. and Freeman, D.J. (2006) 'Pay-As-You-Throw (PAYT) in the US: 2006 Update and Analyse: Final Report'. Online. Available at: www.epa.gov/osw/conserve/tools/payt/pdf/sera06.pdf (accessed 10 February 2012).

Viscusi, W.K., Huber, J. and Bell, J. (2011) 'Promoting Recycling: Private Values, Social Norms, and Economic Incentives', *American Economic Review*, 101: 65–70.

3 Waste generation and delinking

A theoretical model with empirical application to the Italian municipalities[1]

Graziano Abrate and Matteo Ferraris

Introduction

Delinking of economic growth and environmental impact has become an increasingly topical issue across the world and affects regulation and management practices in several industries. The waste management sector is a particularly interesting case: waste volumes are predicted to continue rising unless action is taken to reduce this problem.

There are several measures that can be taken to reduce the environmental impact of refuse. These include production incentives related, for example, to packaging reduction, and consumption incentives, for example, charging tariffs based on volume. Waste collection can be organized so that different types of waste are separated to increase recycling. According to Gellynck and Verhelst (2007), instruments available to public authorities to reduce waste fall into three main groups: pecuniary incentives (e.g. a move from a fixed annual fee for collection of bags to weight-based fees), changes to service levels (availability of kerbside recycling or drop-off recycling centres, and organic waste collection), and incentives to reduce waste. The amounts of waste generated depend also on the community's socio-economic characteristics, such as income and population density.

In Italy, regulation affects both the management structure of the firms operating in the waste sector and waste generation and its prevention. Legislative Decrees 22/1997 (Decreto Ronchi) and 152/2006, introduced a new tariff, TIA (Tariffa Integrata Ambientale), which replaces the old 'TARSU' (Tariffa Ambientale sui Rifiuti Solidi Urbani). However, the TIA has been implemented in only a small share of municipalities, and from 2013 will be replaced by a new municipal tax on waste and waste services – 'TARES' (Tassa sui rifiuti e servizi, Legislative Decree 201/2011). One of the objectives of the TIA was to introduce mechanisms to reward consumers who separated their waste (e.g. into composting, paper, glass, aluminium for recycling) as part of a 'pay-as-you-throw' scheme that provides an incentive for reducing waste and increasing recycling.

In this chapter, we focus on the environmental aspects of refuse generation and collection drawing on the literature on delinking between income and pollution. There is empirical evidence that some pollutants follow an inverse-U-shape relative to income. This is referred to as the 'Environmental Kuznets Curve', or EKC, after Kuznets' (1955) study of the U-shaped relationship between economic growth and inequality.

We develop a specific theoretical model of EKC in the municipal solid waste sector to explain the correlation between income and household waste generation. The theoretical model takes a non-linear form and has been tested empirically. We aim to contribute to work on the connection between theoretical and empirical applications in the field.

Data were collected at the municipality level. The econometric analysis supports the existence of an inverse-U-shaped relation between waste and income and emphasizes the role of other demographic and socio-economic factors. In particular, we investigate the effectiveness of the introduction of TIA as an example of municipal policy aimed at reducing the environmental impact of waste.

This chapter is structured as follows. The next two sections respectively review the theoretical and empirical literature on EKC. Then, we describe the theoretical model used to define amounts of refuse as a function of income and other socio-economic variables. Subsequently, we describe the dataset and discuss the empirical results. Concluding remarks follow.

Theoretical literature

The EKC literature contributes to the debate on delinking (decoupling), understood as the separation of economic activity from increasing pollution.

Economists have proposed several reasons for an income–pollution path: increasing production due to economic growth; structural changes;[2] and increased pressure for improved environmental quality as household income increases. While the first category provides an explanation for a positive income–pollution relationship, the other two categories refer to positive as well as negative relationships (Khanna and Plassmann 2004). For example, if *environmental quality* is a normal good, such that the higher the income the higher is the demand for that good, the structure of preferences will have a direct effect on the economy through market behaviour, and an indirect effect through the pressure on governments to take measures to improve environmental quality (Princen 1999).

Since the concept of EKC is empirically based, theoretical studies in this area are rather scarce and generally try to explain the dynamics in terms of type of technological investments, endogenous spillovers, changing preferences and policy factors. Some studies propose a theoretical underpinning for EKC on the basis of dynamic optimal investment in waste management and waste disposal options.

An interesting theoretical interpretation is the model proposed by Andreoni and Levinson (2001). They develop a static model in which the EKC can be derived directly from the technological link between consumption of a desired good and abatement of its undesirable by-product (generic pollution). An extension of this model uses agent-based theory to analyse the mechanism of EKC (Khanna and Plassmann 2006). Khanna and Plassmann decompose the analysis into two frameworks for the allocation of resources to consumption and environmental effort, and the relationship between specific resources and environmental quality. They propose a simple model that does not assume specific functional forms, allowing its application to a wide range of utility and pollution functions. Consumer preferences can foster or prevent pollution reductions. For example, if, as consumers become richer, they do not exercise enough effort to maintain the environment, the most sophisticated and effective abatement technologies will not prevent pollution from increasing. Conversely, consumers' preferences for being green do not need to be very extreme for pollution ultimately to fall with income if abatement measures are sufficiently effective.

In the literature on the waste sector, Fullerton and Kinnaman (1995a, 1995b) and Podolsky and Spiegel (1998) describe the possibilities for substitution between waste disposal and recycling as part of household waste management. They develop models where households maximize utility subject to a budget constraint, which incorporates a unit price for waste collection. These models underpin a series of solid waste disposal and recycling demand equations.[3]

An interesting strand in the environmental literature studies the costs and benefits of municipal solid waste recycling. Recycling can be considered a private or a public good. The private economic benefit to the consumer is subject to the possibility of reducing the marginal cost of municipal solid waste disposal; the public benefits are preservation of the environment and reduced landfill disposal.

According to Kipperberg (2003), the variables affecting recycling decisions fall into three categories: time, moral motivation and passive-use value. In the first case, recycling is a 'non market good'; thus, to analyse the price of recycling, the cost opportunity of time is used as an explanatory variable. Even if recycling does not imply particular monetary costs for the individual, there might be implicit costs in terms of time spent on recycling. In the case of moral motivations, many consumers have a preference for environmentally friendly behaviour, that is, they want voluntarily to contribute to the public good. Finally, passive-use value is associated, on the one hand, with the eventual economic incentives that consumers may derive from participation in recycling programmes, and on the other hand with a reduction in negative externalities due to increased general levels of recycling. The benefits from environmental preservation are not excludable and are not rivalrous (they increase independently of the recycling behaviour of individual consumers). The heterogeneity of consumers is an important factor since the choice to recycle may in part be voluntary (independent of money incentives) and in part the result of a monetary incentive.

Empirical literature

Empirical verification of the EKC hypothesis is usually not based on a specific theoretical model and there is no consensus on the type of model to use or the variable to include in the analysis.

Most studies adopt linear or log-linear regression. A few make use of non-linear regressions. Berrens *et al.* (1997), using a generalized gamma function to evaluate the empirical parameters in US data, confirm the inverted-U curve hypothesis for hazardous waste. Using a similar empirical methodology, Galeotti and Lanza (1999) identify turning points for different pollutants, by comparing linear and log-linear models to non-linear models (gamma and Weibull functions).

For type of variables, the empirical literature provides numerous applications for different types of pollutants (e.g. carbon monoxide, sulphur dioxide, carbon, suspended particulate matter, ground level ozone, nitrogen oxide, etc.). Some authors argue that, because of the greater local benefits of abatement, local pollution tends to decline with income in middle income level countries, while global pollution continues to increase (Shafik 1994). Khanna and Plassmann (2004) find that the income–pollution relationship turns negative at lower income levels for goods whose consumption can be *spatially separated* from the source of pollution. Where spatial separation is possible, the opportunity cost of pollution abatement is likely to be low and the sign of the equilibrium income elasticity of pollution changes from positive to negative, hence the EKC slopes downwards, at relatively low income levels. Where separation is not possible, the equilibrium income elasticity remains positive up to relatively high income levels. In the following, we focus only on evidence specifically on the waste sector. Table 3.1 presents the main works and the four groups of explanatory variables for level of income/wealth, demographic factors, socio-economic factors and policy decisions.

At the core of delinking analysis is the relationship between refuse generation and development, measured usually as national or provincial GDP or, in some cases, as consumption expenditure. Shafik (1994) was the first empirical study on waste to estimate an increasing monotonic relationship between waste generation and income, thus rejecting the EKC hypothesis. This result is supported by Concu (2000) and Gellynck and Verhelst (2007), who find a positive income elasticity of demand for waste collection services (around 0.326). Johnstone and Labonne (2004) also find that waste generation rates are inelastic with respect to household consumption expenditure per capita, and Karousakis (2006) points only to the presence of a relative delinking between income and waste generation. On the other hand, Mazzanti *et al.* (2008) show evidence of a turning point in the Italian case – although at a high level of income – based on panel data for 1999 to 2005, at the level of province. Cross-country European analyses find evidence of a

Table 3.1 Literature on EKC and the relationships between socio-economic and waste generation variables

Author(s) and publication years	*Data (and source)*	*Model*	*Dependent variables*	*Independent variables*	*EKC and turning point*
Shafik 1994	Panel Data, 149 countries from 1960 to 1990 (World Bank's World Development Report)	Linear Log-linear Quadratic Cubic	MSW*	- GDP p.c.	- Monotonically increasing
Podolsky and Spiegel 1998	149 municipalities in New Jersey		Waste disposal	- Household median age - Household size - Education	- Negative effect - Negative effect - Negative effect
Berrens *et al.* 1998	Cross-section, 3,141 US counties, year 1991 (US EPA – Environmental Protection Agency)	Generalized gamma model	Hazardous waste	- Income p.c.	- Inverted U – shaped EKC (TP: from $17,679 to $20,253)
Fullerton and Kinnaman 1997	OECD		MSW	- Education	- Negative effect
Concu 2000	Cross-section, 322 municipalities of Sardinia	Quadratic, log-linear (arrested at the III terms), Linear (arrested at the III terms)	MSW	- Real estate properties p.c. (in millions of Lire), Dummy Seaside - Touristic specialization	- NO TP for linear and quadratic model YES TP for cubic model - Positive effect
Jenkins *et al.* 2003	Questionnaire to 4,600 households residing in 20 US metropolitan areas	Probit	MSW recycling	- Kerbside programme	- Positive effect

Table 3.1 (cont.)

Author(s) and publication years	*Data (and source)*	*Model*	*Dependent variables*	*Independent variables*	*EKC and turning point*
Johnstone and Labonne 2004	Cross-sectional Time series, 30 UE countries, from 1980 to 2000 (OECD Environment directorate's database)	Linear Log-linear	MSW	- Household income	NO TP, but relative delinking (income positive elasticity between 0.05 and 0.55)
				- Density	- Positive effect
				- Share of children	- Negative effect
				- Urbanization	- Positive effect
Karousakis 2006	Panel data, 30 OECD countries over a period of 30 years (since 1980 to 2000) – 750 observations	Linear	MSW	- GDP	- NO TP, but relative delinking (income elasticity is around 0.42–0.45)
				- Urbanization	- Positive effect
			MSW recycling	- Density	- Negative effect
				- Population growth	- Negative effect
Gellynck and Verhelst 2007	Cross-section, 295 municipalities of Belgium (OVAM on the basis of a structured questionnaire)	Linear	Mixed household MSW	- Income p.c.	- Positive effect
				- Density	- Positive effect
				- Direct costs for service (polluter-pays principle)	- Negative effect
				- Frequency of collection	- Positive effect
				- Organic collection	- Negative effect

Andersen *et al.* 2007	EU15 and EU10 countries	Log-linear	MSW Material flows	- Population - Number of households - Time	Wastes increase at a lower rate than economic growth – Relative decoupling
Mazzanti *et al.* 2008	Panel Data, 103 Italian provinces from 1999 to 2005 (APAT, ISTAT)	Log-linear (arrested to II terms)	MSW	- GDP p.c. (regional) - Density - Separated waste collection - Tariff - Tourism	- TP at Provincial level (from 22.586 to 31,611 euro) - Positive effect - Negative effect - Positive effect - Positive effect
Mazzanti 2008	Panel data, 28 European countries, from 1995 to 2000 for MSW and EU15 from 1997 to 2000 for packaging waste	Log-linear	MSW Packaging waste	- GDP, consumption	- NO bell-shaped curve, no relative or absolute delinking path (elasticity close to one)
Mazzanti and Zoboli 2009	Panel data, EU 25, from 1995 to 2005 (EUROSTAT)	Log-linear (arrested at the II term)	MSW	- Consumption and GDP - Density - Population - Household Size - Age - Policy variable (Landfill and incinerator directive)	- Relative delinking - Positive effect - Positive effect - Positive effect (n.s.) - Positive effect (n.s.) - Negative (n.s.)

Table 3.1 (cont.)

Author(s) and publication years	*Data (and source)*	*Model*	*Dependent variables*	*Independent variables*	*EKC and turning point*
Mazzanti *et al.* 2009	Panel data, 103 Italian provinces of 20 regions, from 2001 to 2007 (APAT)	Log-linear	MSW	- GDP - Density - Tourism - Separated waste - Tariff	No TP, relative delinking (elasticity below unity) - Positive effect - Positive effect - Positive effect - Positive effect (n.s.)

* MSW = Municipal Solid Waste.

relative delinking, but no bell-shaped curve (Andersen *et al.* 2007; Mazzanti and Zoboli 2009; Mazzanti *et al.* 2009).

Among the variables for demographic factors, there is a consensus that population, population density and degree of urbanization are positively correlated to waste generation (Johnstone and Labonne 2004; Gellynck and Verhelst 2007; Mazzanti *et al.* 2008, 2009). Karousakis (2006) notes that, though urbanization is associated with higher levels of waste generation, it has a negative impact on the amount of waste sent to disposal.

Johnstone and Labonne (2004) find evidence that the number of children in the population has a negative and significant effect on waste generation, while Podolsky and Spiegel (1998) show that the higher the median age of the household the lower the amounts of waste generated.

In considering the impact of socio-economic variables, some authors focus on the relationship between education and household waste. Intuition would suggest that better educated households would be more likely to sort their waste and have a higher awareness of environmental and health issues. Podolsky and Spiegel (1998) and Fullerton and Kinnaman (1999) estimate that higher education levels tend to reduce, respectively, the total amount of waste generated and the amounts of non-separated waste.

There is also an impact from tourism. Concu (2000) finds that a tourist industry and protecting the environment are not complementary, that is, tourism is damaging to environmental quality. Studies by Mazzanti *et al.* (2008, 2009) confirm the positive relationship between tourism and waste generation.

The way that local authorities and waste industry managers can influence the behaviour of consumers is estimated by analysing policy variables. Jenkins *et al.* (2003) analyse two solid waste programmes, residential kerbside recycling, and volume-based (or unit) pricing, in 20 metropolitan statistical areas in the United States, based on materials recycled. They analyse recycling of glass (bottles), plastic (bottles), aluminium (cans), newspaper, and yard (garden) waste. They find that kerbside recycling, as opposed to mandatory recycling which has no significant effect, has a positive and significant effect on the share of recycling of these five materials; the effect of unit pricing is not clear. Gellynck and Verhelst (2007) suggest that less frequent collection of non-separated waste, implementation of kerbside collection programmes for organic waste, and collection of organic waste would have a significant impact in the amount of refuse generated. They indicate also that pecuniary incentives are effective instruments and that the higher the amount of the direct costs for waste services that is borne by households, the lower are the amounts of waste collected. This result is confirmed for Italy where Mazzanti *et al.* (2008, 2009) find that separated waste collections tend to decrease waste generation.

Theoretical model

Following Andreoni and Levinson (2001), we consider the problem of utility maximization by a representative household, in order to adapt the EKC theory to the waste sector (for a longer discussion of the theoretical derivation of EKC see also Ferraris 2010).

Let us consider a utility function $U_j(C_j, R_{ij})$, where j is the representative household in the *j-th* municipality (for $j = 1$ to N), such that it is increasing in private good consumption and decreasing in refuse. Refuse generation and disposal (especially in landfill) is considered a 'public bad' and can be interpreted as the degree of environmental damage that generates disutility to the individual consumer, similar to the generic pollutant in Andreoni and Levinson's (2001) model. In our model, we provide two alternative hypotheses: that the public bad is represented either by 'total waste' or 'not-separated' waste, thus $i = TOT, NS$.

Let us consider the household utility function as a function of consumption and refuse:

$$U_j(C_j, R_{ij}) = C_j - zR_{ij} \tag{3.1}$$

where $z > 0$ represents the constant marginal disutility of refuse.

Since refuse is considered a pollutant and a byproduct of consumption, the greater the household's consumption the higher will be the level of pollution. However, let us suppose that the consumer can expend resources to alleviate pollution, that is, to increase recycling or reduce total waste. We call the *environmental effort* (E) the level of involvement in activities aimed at improving environmental quality.

We will impose that refuse generation is given by the following function, decreasing in effort:

$$R_{ij} = f(C_j, E_{ij}) = C_j - C_j^{\alpha} E_{ij}^{\beta} \tag{3.2}$$

Equation (3.2) shows that the positive consumption–waste relationship can be mitigated by the amount of resources spent on environmental effort, which promote a reduction in waste generation. The volume of refuse reduction is described as the output of a Cobb-Douglas technology. Technically, the C_j^{α} variable represents the part $(0 \leq \alpha \leq 1)$ of consumption that can, at least potentially, be delinked from waste with a sufficient level of environmental effort. Thus, if α is equal to zero, the effort will never be productive and no delinking will occur. For example, if our public bad is represented by the non-separated waste, the C_j^{α} variable represents the portion of consumption that can be recycled, thus if α is equal to 1, all consumption potentially can be recycled.

The representative household maximizes its utility function as follows:

$$MaxU_j(C_j, R_{ij}) = Max\left\{C_j - z\left\lfloor C_j - C_J^{\alpha} E_{ij}^{\beta} \right\rfloor\right\}$$
$$s.t. \tag{3.3}$$
$$W_J = p_{c_j} C_j + p_{ij} E_{ij}$$

where W is the income of a representative household (in a municipality), p_c is the price of consumption, and p_E is the price of effort. For simplicity, we can rewrite the budget constraint as:

$$M_j = C_j + \pi_{ij} E_{ij}$$

where $M_j = \frac{W_j}{p_{cj}}$ is the real income and $\pi_{ij} = \frac{p_{E_{ij}}}{p_{C_j}}$ is the real price of the effort.

If we assume that $z = 1$, the first order conditions can be written as:

$$E_{ij}^{*} = \frac{\beta}{\alpha + \beta} \frac{M_j}{\pi_{ij}} \tag{3.4}$$

and

$$C_j^{*} = \frac{\alpha}{\alpha + \beta} M_j \tag{3.5}$$

If we insert the solutions of equations (3.4) and (3.5) into equation (3.2) and compute the first derivatives, we get the 'optimal refuse function', which depends on the technology, the relative prices of the effort, and the income:

$$R_{ij}^{*} = \frac{\alpha}{\alpha + \beta} M_j - \left(\frac{\alpha}{\alpha + \beta}\right)^{\alpha} \left(\frac{\alpha}{\alpha + \beta}\right)^{\beta} \frac{M_j^{\alpha + \beta}}{\pi_{ij}^{\beta}} \tag{3.6}$$

The income path changes according to the values of the parameters. If $\alpha + \beta > 1$ the refuse function is concave, confirming the inverse-U-shaped path described by the EKC.

In the case that $\alpha + \beta > 1$, the turning point (τ^{*}), which represents the maximum amount of income over which refuse generation begins to decrease, is computed by the first order conditions of the function in equation (3.6):

$$\tau^* = (\alpha+\beta)\left[\frac{\alpha^{1-\alpha}}{\alpha+\beta}\left(\frac{\pi_{i,j}}{\beta}\right)^{\beta}\right]^{\frac{1}{\alpha+\beta-1}} \tag{3.7}$$

Equation (3.7) shows that the turning point is also a function of the (relative) price of effort, π. In particular, an increase in the relative price would modify the EKC curvature, pushing moving the turning point to a higher level.

Let us assume S_j is a set of socio-economic, political and environmental variables for each household, municipality or other aggregate. Then, the value of π can be written as a function of these variables and the resulting formulation can be tested empirically:

$$\pi_{ij} \equiv f(S_j) \tag{3.8}$$

In line with the idea of heterogeneity of consumer preferences and habits, the intuition is that the price of effort will vary with the individual and community levels and may depend also on socio-economic and demographic factors. For instance, level of education, lifestyle, household composition (number of family members, average age, etc.), demographic background, population density, and regional environmental policies such as recycling programmes (e.g. volume-based pricing, unit pricing) can raise or lower the cost of separated refuse. Thus, the shape of the EKC and its turning point may be different for each community.

Empirical analysis

Data and variables

The database includes 540 Italian municipalities, observed over the period 2004 to 2006. Publicly available ISTAT data are used as demographic and socio-economic indicators; information on amounts of collected waste are from EcoCerved.[4] The total number of observations is equal to 1,565; missing data means that the panel is not strictly balanced. The sample is geographically well distributed over the national area, and covers 169 municipalities in the northern regions, 153 in the centre and 218 in the south. The sample represents over a third of the Italian population.

Table 3.2 presents the summary statistics of the variables used in the empirical analysis. Per capita production of waste (R_{TOT}) is some 470 kg per year on average (almost 1.3 kg per day), with the share of separated waste collection around 20 per cent. There is quite high variability in the sample in relation to total amount of waste collected and waste sorting policies. While some municipalities do not have a serious recycling programme (more than 100 municipalities – mainly in the south – register less than 5 per cent of

Table 3.2 Summary statistics

Variable description		*Mean*	*Standard deviation*	*Min*	*Max*
Total amount of solid waste collected (kg/year per-capita)	R_{TOT}	468.72	143.23	138.65	1,427.06
Total amount of non-separated waste collected (kg/year per-capita)	R_{NS}	383.73	153.04	60.36	1,409.64
Share of separated collection (%)		19.55	17.73	0.05	76.49
Per-capita real income (€)	M	9,528	3,300	3,143	20,670
Population	S_{POP}	40,993	143,254	992	271,149,1
Population density	S_{DENS}	890	1,237	23	9,442
Household size	S_{HS}	2.56	0.28	1.93	3.56
Tourism	S_{TUR}	6,857	21,721	0	330,581
Share of adult population with high-school degree (%)	S_{ED}	31.92	7.41	14.86	53.96
Presence of incentive-based tariff (Yes=1, No=0)	S_{TIA}	0.25	0.44	0	1

separated waste collected in 2006), others have achieved as much as 76.5 per cent of waste to recycling.

Per capita municipality income (M) is expressed in real term (base = year 2006) and is based on fiscal data from the Ministry of Internal Affairs.[5] Average income in the sample is less than €10,000. Note that this measure of personal income (which is based on personal income tax) is significantly lower than GDP, a common measure in other EKC empirical studies, but not available at the municipality level.

The other variables capture spatial characteristics that are expected to influence the EKC relation through Equation (3.8), thus explaining the price of the effort to reduce waste. First, we account for demographic characteristics such as the municipality size, measured as number in the population (S_{POP}), population density (S_{DENS}) and average household size (S_{HS}). As highlighted in Podolsky and Spiegel (1998), household size can be expected to have an effect on per capita waste due to domestic economies. Second, being a tourist venue can have a significant impact on per capita waste production, since this is computed over the residential population. In the absence of reliable data on numbers of tourists, this effect is measured using a supply-side indicator, that is, total number of beds provided in tourist accommodation per 100,000 inhabitants (S_{TUR}). Third, the effect of education has been noted (S_{ED}). We use share of the adult population with a high-school degree (on average, equal to 32 per cent) to proxy for this effect. We can expect education to play a role in reducing the environmental impact of waste as in Judge and Becker (1993) and Fullerton and Kinnaman (1997). On the other hand, since the education level is correlated to income, it might be mitigated by the opportunity cost of the time required for recycling activity (Kipperberg 2003). Finally, in the context of regulation of the Italian sector, municipalities are supposed to have implemented the TIA which, at least in principle, provides some mechanisms to reward individuals who separate their refuse. We create a dummy variable (S_{TIA}) that takes the value 1 if the TIA has been adopted – around 25 per cent of observations. According to the theoretical model, we would expect the presence of this monetary incentive to compensate for the price of the effort to recycle.

Results

We estimate the EKC relationship by means of non-linear least squares, using the statistical package STATA (version 10.0); the significance of the parameters is tested at different levels ($p<0.01$; $p<0.05$; $p<0.10$). A common issue related to non-linear least squares is model convergence to local maxima, thus the assignment of appropriate initial values to the parameters is crucial (Ratkowsky 1990). In order to circumvent these problems, which increase with the number of parameters to be estimated, we simplify the model by imposing $\alpha = 1$. Given the interpretation of this parameter, the assumption is that all consumption can be recycled. Virtuous cases of municipalities with shares of collected waste of almost 80 per cent support such simplification.[6] We assume also that the relative price of the effort (π) can be expressed as a linear function of the explanatory variables, as follows:

$$p_{ij} = \pi_i + \sum_k \gamma_{ik} S_{ijk} \tag{3.9}$$

S_k represents the set of municipality characteristics and includes two macro-regional dummies, with

$$k = \{DENS, POP, HS, TUR, ED, TIA,\\ NORTH, SOUTH\}.$$

Table 3.3 presents the findings from two alternative models of EKC, which, respectively, consider R_{NS} and R_{TOT} as the dependent variable. In general, the fit of both models appears good: apart from the high value of R-squared, most estimated parameters are highly significant and the predicted values are strongly correlated to the real data.

Table 3.3 Non-linear regression results (standard error in brackets)

	Dependent variable: R_{NS}	*Dependent variable:* R_{TOT}
Number of observations	1,565	1,565
R-squared [Adjusted R-squared]	0.9324 [0.9320]	0.9485 [0.9482]
Correlation of fitted values with real data	0.7136	0.6327
α	1	1
β	0.05070*** (0.00212)	0.04530*** (0.00217)
Determinants of PE		
δ	2,075.039*** (106.9778)	2,664.450*** (156.5146)
γ_{DENS}	−0.00322 (0.00538)	−0.01220 (0.00766)
γ_{POP}	0.00014*** (0.00005)	0.00013** (0.00006)
γ_{HS}	−422.1473*** (38.59946)	−487.7318*** (54.30068)
γ_{TUR}	0.01447*** (0.00123)	0.02293*** (0.0024)
γ_{ED}	5.55730*** (1.04271)	0.42295 (1.56972)
γ_{TIA}	−30.59858** (13.94553)	−11.16346 (20.19206)
γ_{NORTH}	−277.8862*** (20.73175)	−293.9845*** (31.33459)
γ_{SOUTH}	66.15438* (35.42509)	−7.13180 (45.902)

In relation to the determinants of the effort price, the results indicate that heterogeneity across the sample significantly affects the possibility of the municipality reaching a turning point. Among demographic factors, we find a positive impact of municipality size and a negative impact of household size; the density effect is not significant. This means that the largest municipalities tend to find it more difficult to promote waste reduction and waste recycling. Also, the effort involved in reducing waste (and non-separated waste) is less costly for larger households, which is evidence of 'domestic' scale economies. As expected, touristic flows have a significant and positive impact on the price of the effort and, as a consequence, on the turning point.

For the effect of education, we find an important difference between the two EKC models. In the case of total amount of waste produced, there is no significant impact of level of education. Instead, in the model with amount

Table 3.4 Evidence of a turning point

Computation at the sample mean (hypothetical observation)	*Turning point for R_{NS}*			*Turning point for R_{TOT}*		
	9,952			*13,719*		
Computation point-by-point	*Mean*	*S.d.*	*%TP*[a]	*Mean*	*S.d.*	*%TP*[a]
Total sample	9,422	2,847	46.96	12,871	4,689	31.37
North	8,002	1,796	97.10	11,300	2,949	87.16
Centre	10,521	2,330	56.15	14,505	3,943	15.66
South	9,730	3,353	2.36	12,916	5,738	0.00

Note: [a] %TP indicates the share of municipalities that reached the turning point.

of non-separated refuse as the dependent variable, level of education has a significant and positive impact on the price of the effort, supporting the idea that education increases the opportunity cost of the time devoted to the activity of recycling. Similarly, the implementation of the TIA has a significant impact on waste separation, by compensating for the household opportunity cost associated with the recycling effort. However, the incentive is not significant for reducing total waste. This evidence is coherent with the idea that incentives can increase separated waste among consumers, but to reduce waste policy needs to take account also of reductions in the production and packaging phases.

We computed the turning point for a hypothetical municipality with the same demographic and socio-economic characteristics as the average in the sample. The estimates show the presence of a turning point within the range of the observed income for both total amount of solid waste collected and non-separated waste (Table 3.4). In the case of total waste, the value of €13,719 falls in the upper decile of the income distribution, indicating that few municipalities have reached a point where the increasing trend in total waste generation has flattened out. This result is in line with Mazzanti *et al.* (2008), who find evidence of an EKC at the provincial level. The results for non-separated waste are more encouraging: it starts to decrease for incomes slightly higher than the sample average (€9,952). We can visualize the EKC as being decomposed into two environmental 'delinking' levels: at a certain level of income the amount of non-separated waste starts decreasing because it is substituted by separated waste; a substantial additional level of income is required to register a reduction in waste generation (Figure 3.1).

Finally, we account for heterogeneity among socio-economic and demographic characteristics, for each municipality comparing the income

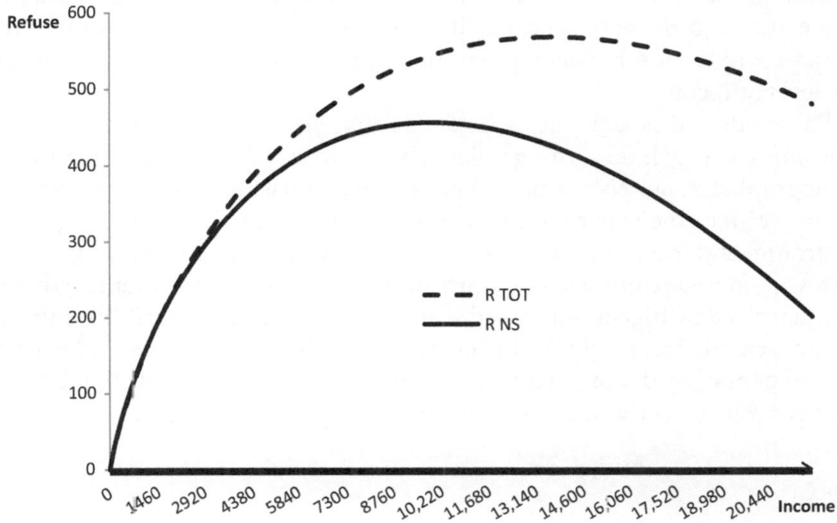

Figure 3.1 Estimated EKC for total and not separated refuse ('average' municipality).

corresponding to the estimated turning point with actual real income. This highlights the substantial regional differences, which are evident also in the estimates for the macro-regional dummies. In particular, there is a large gap between the North of Italy and the rest of the country. Among the whole sample, almost a third of municipalities has reached the turning point for total waste generation; for cities in the north, the proportion is 87 per cent, for the centre it is around 15 per cent, and in the south is zero. For recycling activity, the share is 97 per cent in the north, 56 per cent in the centre, and 2 per cent in the south.

Conclusions

This chapter examined the EKC hypothesis applied to the municipal solid waste sector. We highlighted the importance of recycling for improving environmental quality, verifying the patterns of two alternative indicators of 'pollution': amount of total waste and amount of non-separated waste. Building on the theory proposed by Andreoni and Levinson (2001), we develop a theory highlighting a waste–income relationship that follows an inverse-U-shaped path, depending on the environmental efforts made by households to reduce waste generation and improve waste recycling.

The model takes a non-linear form, which is tested econometrically using data from 540 Italian municipalities for the years 2004–6. We find two environmental 'delinking' levels: with a certain level of income – corresponding to around the median value for the sample – the amount of non-separated waste starts to decrease because it is substituted by separated waste. To achieve a reduction in waste generation requires a substantial addition to the level of income.

The model takes account of heterogeneity, allowing us to derive a specific municipality level turning point. The existence of delinking depends on demographic, socio-economic and policy variables which affect the shape of the curve since they have an impact on the perceived costs of implementing environmental friendly actions. In particular, the turning point depends positively on municipality size and tourism flows and is lower in municipalities characterized by higher household size. The introduction of tariff incentives for households has a significant impact on promoting separated collection by compensating the opportunity cost associated with the recycling effort. However, the incentive is not significant for reducing total waste.

Notes

1 Financial support from the project ICT4LAW (Regione Piemonte) and from the HERMES Research Centre (Turin) is gratefully acknowledged. We thank Giovanni Fraquelli and Giampaolo Vitali for helpful comments and suggestions. The usual disclaimers apply.

2 That is, changes in the economy's output mix or deriving from national economic growth, such as the transformation from an agricultural to an industrial economy, and then a change to dominance of services, which have some positive effects on the environment. The first change is likely to generate increased pollution while the second may result in pollution reduction. In other words, there is an effect deriving from the adoption of techniques and technologies that require fewer resources and generate less pollution per unit of output. In line with this, free trade theory suggests that developing countries will specialize in the production of goods whose inputs are in relative abundance domestically: i.e. labour, natural resources. Developed countries will specialize in human capital and manufacturing capital intensive activities. Some of the reduction in environmental degradation levels in the developed countries, and increases in environmental degradation in middle income countries, are reflections of this specialization. Also, environmental regulation in developed countries might encourage relocation of polluting activities in developing countries.

3 These equations include three types of independent variables: characteristics of goods whose consumption generates waste (includes the price of consumption goods and the amount of waste generated per unit of good); a description of the local waste management system (price per unit of waste disposal, a vector of the recycling programme's features); socio-economic factors (household size, income and education). Each material has unique characteristics which might affect the relationship between recycling and the exogenous variables.

4 EcoCerved is an Italian company, among whose responsibilities is to collect municipal data on waste management.

5 Data were accessed from the website of the business newspaper *Il sole 24 ore*, and adjusted to account for regional differences in purchasing power.

6 We ran sensitivity analysis on the parameter, setting it to 0.8 and 0.9. In neither case did the signs of the parameters and the value of the turning point change significantly; of course, this assumption has an impact on the values of single parameters.

References

Andersen, F., Larsen, H., Skovgaard, M., Moll, S. and Isoard, S. (2007) 'A European Model for Waste and Material Flows', *Resources, Conservation and Recycling*, 49: 421–35.

Andreoni, J. and Levinson, A. (2001) 'The Simple Analytics of the Environmental Kuznets Curve', *Journal of Public Economics*, 80: 269–86.

Berrens, R., Bohara, A., Gawande, K. and Wang, P. (1997) 'Testing the Inverted-U Hypothesis for US Hazardous Waste: An Application of the Generalized Gamma Model', *Economic Letters*, 55: 435–40.

Concu, N. (2000) 'La tirannia del trade-off sconfitta? Turismo, ambiente naturale e rifiuti solidi urbani: la ricerca di una Environmental Kuznets Curve', *Working Paper, CRENoS, Centre for North South Research.*

Ferraris, M. (2010) 'L'Environmental Kuznets Curve nel settore dei rifiuti solidi urbani', *Economia Pubblica*, 1–6: 229–54.

Fullerton, D. and Kinnaman, T.C. (1995a) 'How a Fee per-unit Garbage Affects Aggregate Recycling in a Model with Heterogeneous Households', in A.L. Bovenberg and S. Cnossen (eds), *Public Economics and the Environment in an Imperfect World*, Dordrecht: Kluwer Academic Publishers, pp. 135–59.

Fullerton, D. and Kinnaman, T.C. (1995b) 'Garbage, Recycling and Illicit Burning or Dumping', *Journal of Environmental Economics and Management*, 29: 78–91.

Fullerton, D. and Kinnaman, T.C. (1997) 'Garbage and Recycling in Communities with Curbside Recycling and Unit-Based Pricing', *Working Paper no. 6021*, Washington, DC: National Bureau of Economic Research.

Fullerton, D. and Kinnaman, T.C. (1999) 'The Economics of Residential Solid Waste Management', NBER *Working Paper no. 7326.*

Galeotti, M. and Lanza, A. (1999) 'Desperately Seeking (Environmental) Kuznets', *Working Paper no. 2.99*, Fondazione Eni Enrico Mattei, Milan.

Gellynck, X. and Verhelst, P. (2007) 'Assessing Instruments for Mixed Household Solid Waste Collection Services in the Flemish Region of Belgium', *Resources Conservation and Recycling*, 49: 372–87.

Jenkins, R.R., Martinez, S.A., Palmer, K. and Podolsky, M.J. (2003) 'The Determinants of Household Recycling: A Material-specific Analysis of Recycling Program Features and Unit Pricing', *Journal of Environmental Economics and Management*, 45: 294–318.

Johnstone, N. and Labonne, J. (2004) 'Generation of Household Solid Waste in OECD Countries: An Empirical Analysis Using Macroeconomic Data', *Land Economics*, 80: 529–38.

Judge, R. and Becker, A. (1993) 'Motivating Recycling: A Marginal Cost Analysis', *Contemporary Policy Issues*, 11: 58–68.

Karousakis, K. (2006) 'Municipal Solid Waste Generation, Disposal and Recycling: A Note on OECD Inter-Country Differences', *Working Paper*, School of Public Policy, University College London.

Khanna, N. and Plassmann, F. (2004) 'The Demand for Environmental Quality and the Environmental Kuznets Curve Hypothesis', *Ecological Economics*, 51: 225–36.

Khanna, N. and Plassmann, F. (2006) 'Preferences, Technology and the Environment: Understanding the Environmental Kuznets Curve', *American Journal of Agricultural Economics*, 88: 632–43.

Kipperberg, G. (2003) 'An Investigation into Environmentally Friendly Consumer Behavior: The Case of Household Recycling', Department of Agricultural and Resource Economics, Dissertation Prospectus.

Kuznets, S. (1955) 'Economic Growth and Income Inequality', *The American Economic Review*, 45: 1–28.

Mazzanti, M. (2008) 'Is Waste Generation De-linking from Economic Growth? Empirical Evidence for Europe', *Applied Economic Letters*, 15: 287–91.

Mazzanti, M. and Zoboli, R. (2009) 'Municipal Waste Kuznets Curves: Evidence on Socio-Economic Drivers and Policy Effectiveness from the EU', *Environment and Resource Economics*, 44: 203–230.

Mazzanti, M., Montini, A. and Nicolli, F. (2009) 'Waste Kuznets Curve, Regional Policies and Spatial Effects: Evidence on Waste Generation and Landfill Diversion from Italy', 8th International Conference of the European Society for Ecological Economics, Lubliana, 30 June.

Mazzanti, M., Montini, A. and Zoboli, R. (2008) 'Municipal Waste Generation and Socio Economic Drivers: Evidence from Comparing Northern and Southern Italy', *The Journal of Environment & Development*, 17 (1): 51–69.

Podolsky, M.J. and Spiegel, M. (1998) 'Municipal Waste Disposal: Unit Pricing and Recycling Opportunities', *Public Works Management Policy*, 3: 27–39.

Princen, T. (1999) 'Consumption and Environment: Some Conceptual Issues', *Ecological Economics*, 31: 347–63.

Ratkowsky, D.A. (1990) *Handbook of Non-linear Regression Models*, New York: Dekker.

Shafik, N. (1994) 'Economic Development and Environmental Quality: An Econometric Analysis', *Oxford Economic Papers*, 46: 757–73.

4 A note on illegal waste disposal, corruption and enforcement

Alessio D'Amato and Mariangela Zoli

Introduction

Corruption reduces the effectiveness of economic policies in several areas, including waste management, to which the anecdotal evidence related to the waste crisis in the southern Italian regions is witness (see, among others, D'Alisa *et al.* 2010; Pasotti 2010).

In the case of Italy, there is a large body of evidence showing that the producers of waste, corrupt local authorities and supervisory bodies often interact giving rise to sophisticated illegal systems for falsifying transport and processing documents. The operations of these illegal networks are undermining the correct management of waste collection and disposal (Legambiente 2010, 2011, 2012; Massari and Monzini 2004).

Police investigations show that cooperation among dishonest actors at different stages in the waste cycle is resulting in inventive ways of circumventing the controls and enabling illegal dumping of waste. A popular illegal behaviour is known as 'giro bolla' or 'invoice switch', which involves modifying waste documents so that the material is downgraded from hazardous to non-hazardous waste despite not being appropriately treated. Falsifying documents is made possible by collusive practices involving conniving laboratories that provide false analyses, and cooperative local officers at storage centres. The result is that toxic and hazardous waste simply disappears, and is disposed of as common non-hazardous material.

It has been well documented that widespread corruption in local administrative authorities responsible for monitoring the entire hazardous waste management cycle plays a crucial role in facilitating illegal practices and ensuring impunity for criminal entrepreneurs (Massari and Monzini 2004). Also, waste management is recognized as one of the public administration sectors that is particularly exposed to the risk of connections between corruption and organized crime, especially at the local level and in the southern Italian regions (High Commissioner against Corruption 2006), where mafia-type organizations have a strong hold on society and social institutions, including local public officials.

However, the role of corruption and its effects on the proper functioning of the waste cycle are not limited to specific regions or countries. As Liddick (2010) notes, by the late twentieth century, as tighter waste regulation increased the costs of legal disposal, organized transnational criminals, corporate polluters, and corrupt public officials engaged in an illicit traffic whose scope and profitability are rivalling the international drugs trade. Corruption is exacerbating the problems related to the diffusion of illegal waste trafficking in both developed and developing countries. In several cases, illegal waste disposal is enabled by conniving local authorities, and also corrupt practices at high levels of government. For example, Liddick (2011) reports the case of 15,000 tons of toxic incinerator ash produced in Philadelphia and dumped in Guinea by a Norwegian firm, having been (mis)-labelled as material for brick making. The Norwegian Consul General and other government officials were implicated in this deal (Liddick 2011).

Despite such evidence, the role played by corruptible bureaucrats in controlling illegal waste disposal and its effect on waste policy efficiency have remained largely unexplored. This chapter provides some insights into the complex relationships between illegal disposal, waste policy, corruption and public enforcement. We develop a simple theoretical model where an economic agent (a firm) chooses the level of illegal disposal of the waste related to its production process, taking into account that legal disposal requires payment of a waste tax. Illegal waste can be detected by the local police force; when illegal disposal is exposed, the crime is reported to the public authority (which we call the 'bureaucrat' in the following), which (if 'honest') can decide to report the firm's illegal disposal to a higher level regulator or (if 'corrupt') can protect the polluting firm in exchange for a bribe. The choice to be corrupt will be affected by the bribe offered by the regulated firm and by the probability of the bureaucrat being discovered and punished for bribery, which will depend on the enforcement effort exerted by the central government.

In our model, we keep waste related taxation and enforcement efforts (by the local and central police authorities) exogenous, since our interest is in their impact on corruption choices and illegal disposal rather than on their drivers.

In our simplified setting, quite interestingly, a stricter waste policy (i.e. an increase in the charges for legal disposal) increases the probability of corruption, leading to higher levels of illegal disposal and a larger bribe. This is evidence of a trade-off between reducing corruption and financing legal disposal costs. Also, enforcement on the bureaucrat reduces illegal disposal and the probability of corruption (as expected), while, surprisingly, firm-related enforcement can increase illegal waste and bring about a larger probability of corruption.

This chapter is organized as follows. The next section provides an overview of the literature related to waste policy, illegal behaviours and corruption. Our theoretical model is then introduced and main results are derived. Some final remarks conclude our work.

Related literature

The rationales for this work and the basic features of its modelling framework originate in several contributions in various strands of the literature.

A first strand of research is the theoretical work on waste policy in the presence of illegal dumping, starting with Sullivan (1987). In particular, Fullerton and Kinnaman (1995) analyse optimal waste policy in a general equilibrium setting, under the assumption that illicit burning or dumping cannot be taxed directly. In this setting, the authors conclude that first best can be achieved even in the presence of illegal disposal and that the optimal fee structure is a deposit-refund system: a tax on all output plus a rebate on proper disposal in the form of recycling or garbage collection. Following these initial contributions, Choe and Fraser (1999) identify the second-best optimal policy by explicitly introducing monitoring costs into the model, whereas Shinkuma (2003) also accounts for transaction costs. D'Amato and Zoli (2012) investigate the role of organized crime in waste management, explicitly modelling a criminal organization which extorts (socially costly) rent from agents willing to perform illegal disposal. Their main findings suggest that, under certain conditions, a mafia presence can lead to increased levels of economic activity and lower levels of enforcement; in some cases, the related benefits may offset the damages from increased illegal disposal and the social costs of mafia rents.

We refer also to contributions on corruption of officials in order to evade regulation (see the pioneering article by Becker and Stigler 1974), and on the idea developed in Mookherjee and Png (1995), on which the present work is based, of a regulator that delegates environmental enforcement to an inspector responsible for monitoring the pollution emitted by a firm. This action provides scope for corruption because the firm can bribe the inspector to under-report its pollution levels. Bribery is possible because the regulator can neither control the inspector's monitoring effort nor prevent the firm from bribing the inspector. However, the regulator can try to increase the inspector's monitoring efforts by offering a reward (a commission on the fines collected from the firm for pollution). Further, in the case of bribery being discovered, the regulator can penalize both the inspector and the firm. The reward rate, the penalty imposed on the firm giving the bribe and the penalty extracted from the inspector accepting the bribe constitute the three policy instruments available to the regulator. Given the regulator's policies, the firm chooses its pollution level and the inspector chooses his/her monitoring intensity simultaneously.

From this perspective, the authors examine the trade-off between reducing corruption and pollution, showing that when bribery is profitable, increasing the penalties for corruption unambiguously reduces the inspector's incentive to monitor whilst pollution may either rise or fall. The ambiguous result for pollution depends on the existence of different interactions between penalties, monitoring efforts and pollution choices. A small increase in the penalty rate applied to the inspector, for instance, leads to a demand for a larger bribe (to compensate in part for the expected increased penalty) and a less stringent monitoring effort. These two effects have contrasting consequences for pollution (i.e. the first increases the price of pollution while the second reduces it). Analogously, a small increase in the penalty imposed for bribing reduces the size of the bribe provided by the firm and reduces the inspector's incentive to monitor. For the firm, a higher penalty outweighs the smaller bribe and leads to lower levels of pollution, but the reduction in monitoring tends to reduce the expected penalty and leads to higher levels of pollution.

At the opposite, a small change to the inspector's compensation policy unambiguously reduces pollution, increases the size of the bribe and, hence, the inspector's incentive to monitor. In equilibrium, however, a reduction in pollution tends to discourage monitoring efforts, whose intensity finally may rise or fall. Although bribery encourages monitoring efforts, Mookherjee and Png (1995) conclude that corruption is an inefficient way to reduce pollution when the regulator can identify whether the law enforcers are failing to report pollution because they are corrupt or because they cannot observe the pollution. However, when the regulator cannot distinguish legitimate from illegitimate under-reporting, it may be optimal to tolerate some corruption (eradicating bribery would entail excessive costs).

Another paper that is closely connected to our work is by Polinsky and Shavell (2001), and addresses corruption (also) in the form of bribery of public enforcers to avoid their reporting violations. A relevant conclusion in their paper is that corruption dilutes deterrence, because the bribe payments cost less than the sanctions. Also, it may be desirable to monitor and punish corruption in order to raise the offender's costs in terms of bribe payments plus the expected sanction for bribery.

More recent contributions (Samuel 2009; Dechenaux and Samuel 2012) explicitly model the possibility for pre-emptive corruption (i.e. corruption taking place to avoid inspections). In our work we only focus on ex post corruption, that is, bribery takes place after the illegal act has been discovered.

Another interesting perspective on the effects of corruption is provided by some recent studies that explore the impact of bribery in efficiency terms.

Dal Bó and Rossi (2007), for instance, investigate the relation between corruption and the efficiency of electricity distributors in Latin America, from a theoretical and an empirical viewpoint. Their theoretical model assumes that the firm's output depends not only on the amount of labour but also on the managers' supervision, which has the effect of raising the firm's productivity. The idea is that, given a number of workers, managers that supervise them closely obtain better performance and higher output. In this context, corruption diverts managerial effort away from the productive process, implying that more inputs are needed to meet the firm's service obligations. The theoretical prediction of the model is that more corrupt countries will be characterized by less efficient firms. They test this prediction in the empirical part of their study and find that more corrupt countries tend to have more inefficient firms, that is, firms that employ more labour to produce a given level of output. They find also that public firms are less efficient than private firms and that the estimated effects are large in economic terms.

The link between corruption and efficiency is investigated also by Abrate *et al.* (2012), with specific reference to solid waste collection activities in Italian municipalities. By adopting a stochastic latent class cost frontier approach, the authors show that corruption significantly increases cost inefficiencies in the provision of waste services by municipalities. This result is shown to be robust to different specifications of corruption and to the inclusion of different types of control variables. Further, the negative impact of corruption tends to be stronger in southern Italian regions and in municipalities where recycling activities are less developed. Drawing on these results, Abrate *et al.* (2012) suggest that fighting corruption and organized crime can bring significant efficiency gains (calculated to be in the order of €10–50 million for two of Italy's biggest cities, Milan and Rome).

The theoretical model and main results

The theoretical framework proposed in this chapter models a 'representative' agent (e.g. a firm), which performs an economic activity (e.g. production), that generates some waste. The total amount of this activity is assumed to be exogenous and is denoted by y. A local public authority (the bureaucrat) is charged by a higher level authority with enforcing waste related regulation, represented by a unit tax (t) on legal disposal; real-life examples of the latter include landfill tax and waste charges. To legally dispose of waste, the regulated agent has to pay the tax, in addition to strictly convex costs of waste disposal. However, by illegally dumping its waste, the agent can avoid all forms of legal disposal costs. The amount of illegal disposal is denoted by b, so that, given the overall level of economic activity, the amount of legal disposal is given by $y - b$. Convex legal disposal costs by the regulated agent are given by $\frac{\mu(y-b)^2}{2}$, where $\mu > 0$.

In our setting, enforcement is modelled very simply which allows us to distinguish the impact of changes in enforcement efforts on the probability of corruption and on the amount of illegal waste. To this end, we assume that the bureaucrat can detect illegal disposal with a certain probability, $\gamma \in (0, 1)$, which is determined by the monitoring effort of a local public police force. This probability is expected to depend on a number of factors (i.e. the available budget, the number of police officers, etc.) that are outside the bureaucrat's control. In other words, we assume that the probability of detection cannot be strategically manipulated by the bureaucrat in order to increase its expected benefits from bribery.[1]

If illegal disposal is observed by the bureaucrat, the firm may try to corrupt the bureaucrat by offering a bribe (labelled as m). If corruption is present, the bureaucrat does not communicate the presence of illegal disposal to the central authority (more precisely, the bureaucrat communicates a level of b equal to zero); otherwise, the bureaucrat reports the true level of illegal disposal (b) and corruption does not occur.[2]

The bureaucrat, in its turn, is subject to enforcement by the national authority responsible for fighting corruption. The probability that the bribery will be discovered (when it does take place) depends on the enforcement effort exerted by the national authority, summarized in a parameter $v \in (0, 1)$. In the event of the corruption being discovered, the national authority also uncovers the illegal disposal b and both firm and bureaucrat must pay a unit fine (equal to F). Further, if corruption is discovered, the bribe is sequestered.

If corruption does not occur (i.e. the honest bureaucrat correctly reports illegal dumping), the firm involved in illegal disposal must pay a fine X on each unit of illegal waste.[3] The unit fine X is, in our setting, a measure of enforcement directly devoted to the regulated agent. Note again that local enforcement efforts directed towards the firm (to detect illegal disposal) and central enforcement efforts directed towards the bureaucrat (to reveal corruption) are intentionally exogenous in our model because, as explained in the introduction, the aim is to identify the impact of exogenous changes to regulatory stringency and monitoring efforts, on corruption incentives and waste related illegal behaviour, rather than to investigate the determinants of optimal waste policy and enforcement.

Although our modelling strategy is similar to Mookherjee and Png (1995), it differs in that, in our model, the probability of corruption depends on the level of illegal disposal and, also, indirectly, on the waste tax. This enables a better analysis of the complex trade-offs between waste management policies and incentives to be corrupt. We show that our results hinge on the impact of a bribe on the probability of corruption, and on the ability of the firm to affect this probability in reaction to changes in enforcement efforts. We also explicitly disentangle enforcement devoted to the bureaucrat from that exerted on the agent directly. As we will clarify below, this may lead to counterintuitive results.

The game proceeds as follows. In the first stage, for a given level of economic activity, the regulated agent chooses the levels of illegal disposal and the bribe in order to minimize the expected costs of waste disposal, anticipating the impact of its choices on the decision of the bureaucrat to be corrupt (see below). In the second stage, the bureaucrat uncovers illegal disposal b with probability γ; when illegal disposal is detected, corruption occurs if the related expected net benefits are larger than those expected from being honest.

The corruption choice

We assume that the bureaucrat's honesty can be summed up *ex ante* by an 'honesty random variable', which is uniformly distributed, that is $h \sim U(0, H)$. H might be the maximum opportunity cost of not being honest, and depends on the personal characteristics of the local public bureaucrat and/or on the social capital embedded in the local community.

We normalize the bureaucrat's salary to 0.[4] This means that the bureaucrat will choose to act corruptly if:

$$(1-v)m - vFb \geq h$$

or, in words, when the expected benefits from bribery (which accounts for potential distraint of the bribe and the expected fine in the case of detection by the national authority) exceed those of being honest. Modelled in this way, it follows that the bureaucrat takes into account the extent of the illegal disposal when deciding whether to be corrupt or not.

The bureaucrat's *ex ante* probability to be corrupt then is defined as:[5]

$$\Phi((1-v)m - vFb) = \frac{(1-v)m - vFb}{H}.$$

This probability is taken into account by the regulated agent when it evaluates expected illegal disposal costs.

The regulated agent's problem

Given the probability of corrupt local enforcers (defined above), the regulated agent chooses the level of illegal disposal and the bribe that will be offered to the bureaucrat in order to minimize its expected costs, given by:

$$\min_{m,b} \frac{\mu(y-b)^2}{2} + t(y-b) + \gamma((1-\Phi(.))Xb + \Phi(.)(m+vFb)). \tag{4.1}$$

The first order conditions imply that, in equilibrium, the level of illegal disposal (b_n) and the bribe (m_n) are respectively:[6]

$$b_n = \frac{2H(1-v)(t - X\gamma + y\mu)}{2\mu(1-v)H - \gamma(X(1-v) + Fv(v-2))^2} \tag{4.2}$$

$$m_n = \frac{\left(X(1-v)+Fv^2\right)(t-X\gamma+y\mu)H}{2\mu(1-v)H-\gamma\left(X(1-v)+Fv(v-2)\right)^2}. \tag{4.3}$$

Comparative statics

A first interesting insight comes from the impact of a change in the tax rate on illegal behaviour and corruption.[7]

Proposition 1 *An increase in the tax on legal disposal implies increases in illegal disposal, in the bribe and, as a result, in the corruption probability.*

This is plausible and suggests the existence of a trade-off between waste disposal policy on the one hand, and illegal behaviour and corruption on the other. The intuition is straightforward: an increase in the waste tax t increases legal disposal costs, leading to a larger amount of illegal disposal b_n. The agent's marginal willingness to corrupt the bureaucrat increases, so that the bribe m_n increases as well, more than counterbalancing the impact of the increase in illegal disposal on the corruption probability.

Turning to the honesty parameter H, we get the following, rather straightforward, result.

Proposition 2 *A larger honesty parameter brings about smaller illegal disposal, bribe and probability of corruption.*

A larger value for H implies a larger range for the honesty random variable, making at the same time the probability of corruption smaller for any given level of illegal disposal and bribe. Under our assumptions guaranteeing a non-negative corruption probability in equilibrium, the regulated agent has an incentive to react to a larger H by decreasing both illegal disposal and the bribe. However, the change in the latter, together with the direct impact of the change in H, always dominate in equilibrium. These conclusions are quite reasonable if we look at H as a measure of the willingness of the bureaucrat to be corrupt.

By considering the enforcement exerted by the national authority on the bureaucrat, we derive the following:

Proposition 3 *A stricter enforcement on the bureaucrat (as measured by a larger v or F) implies a lower level of illegal disposal and corruption. The same holds if the monitoring effort on the regulated agent is increased.*

The comparative statics signs are presented in Table 4.1.

The intuition is clear also in this case: stronger enforcement on the bureaucrat (measured by a higher v or by a higher F) increases the expected costs of corruption and from illegal disposal, so that the latter decreases, *ceteris paribus*. The regulated agent reacts by decreasing the bribe, and such a decrease always dominates (in equilibrium) the impact of the decrease in

Table 4.1 Comparative statics results

$\frac{\partial b_n}{\partial v} < 0$	$\frac{\partial b_n}{\partial F} < 0$	$\frac{\partial b_n}{\partial \gamma} < 0$
$\frac{\partial \Phi(.)}{\partial v} < 0$	$\frac{\partial \Phi(.)}{\partial F} < 0$	$\frac{\partial \Phi(.)}{\partial \gamma} < 0$

illegal disposal. The final outcome is a reduction in the probability of corruption. A similar intuition applies when dealing with the impact of γ on illegal disposal and on the probability of corruption.

The results are less clear-cut if we look at the impact of the unit fine (X) paid by the regulated agent. As already noted, a honest bureaucrat correctly communicates to the central authority the true level of illegal disposal. The impacts of X are summarized by the following proposition:

Proposition 4 *An increase in the unit fine (X), which directly affects the regulated agent, may lead to increased corruption and illegal disposal depending on the level of the upper bound of the honesty parameter (H).*

The result related to the impact of X is quite surprising. An increase in X can indeed bring about an incentive for the regulated agent to act on illegal disposal and the bribe to increase the probability of corruption, as a larger X implies larger unit costs for illegal disposal when the bureaucrat turns out not to be corrupt *ex post*. When H is sufficiently low (i.e. lower than a threshold value H_X, see the Appendix), an increase in X implies an increase in illegal disposal, as in this case the direct impact of a larger X is relatively small. At the same time, we show in the Appendix that $H < H_X$ is also sufficient to have $\frac{\partial \Phi(.)}{\partial X} > 0$.

Concluding remarks

Illegal behaviour and corruption are widespread in the waste sector. The related distortions should be taken seriously by policy makers when designing waste and enforcement policies. This chapter is a first step towards a better understanding of the complex relations between corruption and illegal waste disposal on the one side, and environmental and enforcement policies on the other. Starting with a very simple waste policy model, we assessed the impact of changes in (exogenous) waste policy and enforcement efforts and provided some counterintuitive results that might have an impact on the effectiveness of waste related intervention.

Specifically, we have shown that the waste policy has an adverse effect on illegal means of disposal and corruption and, as expected, enforcement efforts related to corruption tend to reduce illegal waste and the probability of corruption. However, our results suggest also that an increase in the fine imposed on the waste producing firm can lead to higher levels of bribery and illegal disposal, depending on the general level of corruption in the society. Although the analysis in this chapter is meant to be a starting point, it uncovers several issues that seem to contradict 'commonsense' conclusions, and suggest the need for more research in the field. From a theoretical perspective, it would be interesting to extend our framework, allowing for less clear-cut modelling of corruption. However, we would not expect our results to be significantly affected. Some of our theoretical findings could be tested empirically to investigate the role played by country specific features, such as waste policy interventions and other institutional aspects, in confirming (or not) our theoretical insights.

Appendix

Preliminaries

The convexity of (4.1) entails:

$$\mu + \frac{2\gamma Fv}{H}(X - Fv) > 0$$

which is always the case, since for corruption to be possible we must have $X > Fv$, and

$$\begin{vmatrix} \mu + \frac{2\gamma Fv}{H}(X - Fv) & -\frac{1}{H}\left(F\gamma v^2 + X\gamma(1-v)\right) \\ -\frac{1}{H}\left(F\gamma v^2 + X\gamma(1-v)\right) & \frac{2}{H}\gamma(1-v) \end{vmatrix} > 0$$

i.e. the Hessian determinant must be positive. It requires

$$H > H_c = \frac{1}{2}\frac{\gamma\left(X(1-v) + Fv(v-2)\right)^2}{\mu(1-v)}.$$

Also, for our results to make economic sense, we need to ensure that Φ (.) $\in$ (0, 1), which requires:

$$(1-v)m_n - vFb_n > 0 \tag{4.4}$$

and

$$(1-v)m_n - vFb_n - H < 0. \tag{4.5}$$

A positive corruption probability implies:

$$\Phi(.) = (1-v)(t - X\gamma + y\mu)\frac{X(1-v)+Fv(v-2)}{2\mu(1-v)H - \gamma(X(1-v)+Fv(v-2))^2} > 0.$$

We need to assume $t > X\gamma - y\mu$ to guarantee that both illegal disposal and bribe are positive in equilibrium. It means that the above inequality also implies assuming $X > Fv\frac{(2-v)}{(1-v)}$. In order for (4.5) to hold we need to assume $H > H_1$, where $H_1 = \frac{(X(1-v)+Fv(v-2))((1-v)(t-X\gamma+y\mu)+\gamma(X(1-v)+Fv(v-2)))}{2\mu(1-v)} > H_c$.

Proof of proposition 1

From (4.2) and (4.3) we can derive:

$$\frac{\partial b_n}{\partial t} = 2H\frac{1-v}{2\mu(1-v)H - \gamma(X(1-v)+Fv(v-2))^2} > 0,$$

$$\frac{\partial m_n}{\partial t} = H\frac{X(1-v)+Fv^2}{2\mu(1-v)H - \gamma(X(1-v)+Fv(v-2))^2} > 0,$$

and

$$\frac{\partial\Phi(.)}{\partial t} = (1-v)\frac{X(1-v)+Fv(v-2)}{2\mu(1-v)H - \gamma(X(1-v)+Fv(v-2))^2} > 0,$$

where the sign of the inequalities follows from convexity assumptions, and the last inequality also stems from $X > Fv\frac{(2-v)}{(1-v)}$, which guarantees that $\Phi(.) > 0$.

Proof of proposition 2

Differentiating (4.2) and (4.3) with respect to H we get:

$$\frac{\partial b_n}{\partial H} = 2\gamma(v-1)(t - X\gamma + y\mu)$$

$$\times \frac{\left(X(1-v)+Fv(v-2)\right)^2}{\left(-\left(2\mu(1-v)H - \gamma\left(X(1-v)+Fv(v-2)\right)^2\right)\right)^2} < 0,$$

$$\frac{\partial m_n}{\partial H} = -\gamma(t - X\gamma + y\mu)\left(Fv^2 + X(1-v)\right)$$

$$\times \frac{\left(X(1-v)+Fv(v-2)\right)^2}{\left(-\left(2\mu(1-v)H - \gamma\left(X(1-v)+Fv(v-2)\right)^2\right)\right)^2} < 0,$$

$$\frac{\partial \Phi(.)}{\partial H} = -2\mu(v-1)^2(t - X\gamma + y\mu)$$

$$\times \frac{X(1-v)+Fv(v-2)}{\left(-\left(2\mu(1-v)H - \gamma\left(X(1-v)+Fv(v-2)\right)^2\right)\right)^2} < 0.$$

Proof of proposition 3

Comparative statics with respect to v implies:

$$\frac{\partial b_n}{\partial v} =$$

$$= -\Lambda(2H\gamma(F(4-6v+3v^2)+X(1-v))(X(1-v)+Fv(v-2))) < 0,$$

where $\Lambda = \frac{(t - X\gamma + y\mu)}{\left(2\mu(1-v)H - \gamma\left(X(1-v)+Fv(v-2)\right)^2\right)^2} > 0$. The conclusion that $\frac{\partial b_n}{\partial v} < 0$ follows from the assumptions that $t - X\gamma + y\mu > 0$ and $X > Fv\frac{(2-v)}{(1-v)}$, and from $F(4 - 6v + 3v^2) + X(1 - v) > 0$ for $v \in (0, 1)$.

Turning to the impact of v on the corruption probability, we have:

$$\frac{\partial \Phi(.)}{\partial v} =$$

$$= \Lambda(2\mu(1-v)^2(2F(v-1)-X)H +$$

$$-F\gamma(v^2+2(1-v))(X(1-v)+Fv(v-2))^2) < 0$$

because it can easily be shown that $v^2 + 2(1-v) > 0$ for $v \in (0,1)$.

Our assumptions about parameter values lead to the following signs for the derivatives when the unit fine for corruption is considered:

$$\frac{\partial b_n}{\partial F} = -4Hv\gamma(1-v)(2-v)\Lambda\big(X(1-v)+Fv(v-2)\big)<0$$

and

$$\frac{\partial \Phi(.)}{\partial F} = -\Lambda v(1-v)(2-v)\Big(2\mu(1-v)H+\gamma\big(X(1-v)+Fv(v-2)\big)^2\Big)<0.$$

Accordingly, in equilibrium, m_n and b_n move in the same direction in reaction to changes in v and/or F. However, it is clear that the impact related to m_n is stronger than the impact related to b_n, as the probability of corruption decreases.

The comparative statics for γ implies:

$$\frac{\partial b_n}{\partial \gamma} = 2H\frac{(1-v)\Big(2X\mu(v-1)H+\big(X(1-v)+Fv(v-2)\big)^2(t+y\mu)\Big)}{\Big(2\mu(1-v)H-\gamma\big(X(1-v)+Fv(v-2)\big)^2\Big)^2};$$

given our assumptions for convexity, this derivative is indeed positive for $H < H_\gamma = \frac{1}{2X\mu}\frac{t+y\mu}{1-v}\big(X(1-v)+Fv(v-2)\big)^2$. As however

$H_1 - H_\gamma = \frac{1}{2}\frac{F}{X}\frac{v}{\mu(1-v)}(2-v)(t-X\gamma+y\mu)\big(X(1-v)+Fv(v-2)\big)>0$ and as we assume

$H > H_1$ to guarantee that the probability of corruption is lower than 1, then we can conclude that it is always $H > H_\gamma$, so that $\frac{\partial b_n}{\partial \gamma}<0$

Further:

$$\frac{\partial \Phi(.)}{\partial \gamma} = (1-v) \times \frac{\big(X(1-v)+Fv(v-2)\big)\Big(2X\mu(v-1)H+\big(X(1-v)+Fv(v-2)\big)^2(t+y\mu)\Big)}{\Big(2\mu(1-v)H-\gamma\big(X(1-v)+Fv(v-2)\big)^2\Big)^2};$$

as $X > Fv\frac{(2-v)}{(1-v)}$ by assumption, then $\frac{\partial \Phi(.)}{\partial \gamma}<0$ can be obtained following the same reasoning as for b_n.

Proof of proposition 4

Turning to the impact of a change in X, we get:

$$\frac{\partial b_n}{\partial X} = 2H\gamma(v-1) \times \frac{\left(2\mu(1-v)H - (X(1-v)+Fv(v-2))\right)\left(v\gamma(v-2)F + (1-v)(2y\mu + 2t - X\gamma)\right)}{\left(2\mu(1-v)H - \gamma(X(1-v)+Fv(v-2))^2\right)^2};$$

under our assumptions, $X > Fv\frac{(2-v)}{(1-v)}$ and $(t - X\gamma + y\mu) > 0$, so that it is also $2t - X\gamma + 2y\mu > 0$. As a result, $\frac{\partial b_n}{\partial X} > 0$ when:

$$H < H_X = \frac{1}{2\mu}\frac{(1-v)(2t - X\gamma + 2y\mu) - Fv\gamma(2-v)}{1-v}\left(X(1-v) + Fv(v-2)\right).$$

Notice that $H_X - H_1 = \frac{1}{2}\left(X(1-v)+Fv(v-2)\right)\frac{t - X\gamma + y\mu}{\mu} > 0$ so that $\frac{\partial b_n}{\partial X} > 0$ does not violate any of the conditions on H.

The probability of corruption can be rewritten as:

$$\Phi(.) = \Omega(X(1-v) + Fv(v-2))$$

where $\Omega = \frac{(1-v)(t - X\gamma + y\mu)}{2\mu(1-v)H - \gamma(X(1-v)+Fv(v-2))^2} > 0.$

As $(X(1-v)+Fv(v-2))$ is positive and increasing in X, a sufficient condition for the probability of corruption to increase with X is to have $\frac{d\Omega}{dX} > 0$. It is easily shown that

$$\frac{d\Omega}{dX} = \gamma \times \frac{(v-1)\left(2\mu(1-v)H - (X(1-v)+Fv(v-2))\left((1-v)(2t - X\gamma + 2y\mu) - Fv\gamma(2-v)\right)\right)}{\left(-\left(2\mu(1-v)H - \gamma(X(1-v)+Fv(v-2))^2\right)\right)^2}$$

so that $\frac{d\Omega}{dX} > 0$ when $H < H_X$.

Notes

1 The impact of such strategic manipulation on corruption and illegal behaviour has already been investigated by Mookherjee and Png (1995).
2 Notice that simply trying to bribe the bureaucrat is not punished in our model. The consequences of removing this hypothesis could be investigated in further research.
3 A necessary condition for corruption to take place requires that $vF < X$, i.e. the expected fine for corruption must be lower than the fine that would be imposed for illegal dumping in the absence of corruption.
4 Others, such as Mookherjee and Png (1995), explicitly address the impact of changes in the bureaucrat's salary on the likelihood of corruption. We abstract from these analyses.
5 We assume that parameter values guarantee $\Phi((1-v)m - vFb) \in (0,1)$ and $y > b_n$. We restrict our analysis to the case of a sufficiently large H_l (see the Appendix).
6 Strictly positive values for the bribe and the illegal disposal in equilibrium require the additional assumption that $t > X\gamma - y\mu$, i.e. a sufficiently high tax on legal disposal and/or a sufficiently low fine on illegal disposal if the bureaucrat turns out to be honest *ex post*.
7 The Appendix provides details of the comparative statics.

References

Abrate, G., Erbetta, F., Fraquelli, G. and Vannoni, D. (2012) 'The Costs of Corruption in the Italian Solid Waste Industry', Università di Torino, Department of Economics and Statistics Working Paper Series, N.4.

Becker, G.S. and Stigler, G.J. (1974) 'Law Enforcement, Malfeasance and the Compensation of Enforcers', *Journal of Legal Studies*, 3: 1–18.

Choe, C. and Fraser, I. (1999) 'An Economic Analysis of Household Waste Management', *Journal of Environmental Economics and Management*, 38: 234–46.

Dal Bó, E. and Rossi, M.A. (2007) 'Corruption and Inefficiency: Theory and Evidence from Electric Utilities', *Journal of Public Economics*, 91: 939–62.

D'Alisa, G., Burgalassi, D., Healy, H. and Walter, M. (2010) 'Conflict in Campania: Waste Emergency or Crisis of Democracy', *Ecological Economics*, 70: 239–49.

D'Amato, A. and Zoli, M. (2012) 'Illegal Waste Disposal in the Time of the Mafia: A Tale of Enforcement and Social Well Being', *Journal of Environmental Planning and Management*, 55: 637–55.

Dechenaux, E. and Samuel, A. (2012) 'Pre-emptive Corruption, Hold-up and Repeated Interactions', *Economica*, 79: 258–83.

Fullerton, D. and Kinnaman, T.C. (1995) 'Garbage, Recycling, and Illicit Burning or Dumping', *Journal of Environmental Economics and Management*, 29: 78–91.

High Commissioner against Corruption (Alto Commissariato per la prevenzione e il contrasto della corruzione e delle altre forme di illecito nella Pubblica Amministrazione) (2006) *Studio sui pericoli di condizionamento della Pubblica Amministrazione da parte della Criminalità Organizzata*.

Legambiente (2010) *Rapporto Ecomafie 2010*, Edizioni Ambiente.

Legambiente (2011) *Rapporto Ecomafie 2011*, Edizioni Ambiente.

Legambiente (2012) *Rapporto Ecomafie 2012*, Edizioni Ambiente.

Liddick, D. (2010) 'The Traffic in Garbage and Hazardous Wastes: An Overview', *Trends in Organized Crime*, 13: 134–46.

Liddick, D. (2011) *Crimes Against Nature: Illegal Industries and the Global Environment*, Westport: Praeger.

Massari, M. and Monzini, P. (2004) 'Dirty Business in Italy: A Case Study of Trafficking in Hazardous Waste', *Global Crime*, 6: 285–304.

Mookherjee, D. and Png, I. (1995) 'Corruptible Law Enforcers: How Should They Be Compensated?', *Economic Journal*, 105: 145–59.

Pasotti, E. (2010) 'Sorting through the Trash: The Waste Management Crisis in Southern Italy', *South European Society and Politics*, 15: 289–307.

Polinsky, A.M. and Shavell, S. (2001) 'Corruption and Optimal Law Enforcement', *Journal of Public Economics*, 81: 1–24.

Samuel, A. (2009) 'Preemptive Collusion among Corruptible Law Enforcers', *Journal of Economic Behavior and Organization*, 71: 441–50.

Shinkuma, T. (2003) 'On the Second-Best Policy of Household's Waste Recycling', *Environmental and Resource Economics*, 24: 77–95.

Sullivan, A.M. (1987) 'Policy Options for Toxics Disposal: Laissez-Faire, Subsidization, and Enforcement', *Journal of Environmental Economics and Management*, 14: 58–71.

5 Separate collection target

Why 65 per cent in 2012? Campania case study

Giacomo D'Alisa and Maria Federica Di Nola

Introduction

The waste crisis in Campania can be considered an icon of waste mismanagement in Europe. It has inspired a huge number of studies that have tried to unravel the complex problems involved, and which focus on specific aspects of this crisis. These include: the illegal trade of toxic waste between the Camorra and corporations across Italy and Europe (Fontana *et al.* 2008; Iacuelli 2007); the abuse of legal power by government, and the institutional responsibilities of waste mismanagement (Lucarelli 2007a, 2007b, 2007c; Raimondi 2007); the health impacts of landfill disposal facilities (Fazzo *et al.* 2008; Martuzzi *et al.* 2008; Senior and Mazza 2004; Comella 2007); the impacts of landfill sites on hydrogeological stability (de Medici 2007; Ortolani 2008); the criminal and institutional responsibilities in waste mismanagement (Rabitti 2008); the emergence of new forms of environmentalism, and the relations between society and nature in environmental conflicts (Armiero 2008); the role of activism in the Campania case (Musella 2008); the political implications of the crisis (Barbieri and Piglionica 2007), and the anti-democratic direction characterizing waste management in Campania since the early 1990s (D'Alisa *et al.* 2010).

However, very little research has been conducted on waste generation and waste disposal in Campania. Comparative analyses for Italy identify average regional waste patterns, but cannot explain 'specific hot spots ... e.g. the well known case of Naples and Campania region' (Nicolli *et al.* 2011: 12). Drawing on these analyses, D'Alisa *et al.* (2012) argue that conventional indicators, such as total amount of waste generated, waste generated per capita, fraction of separated collection and the waste disposed of, need to be complemented by other factors to produce a complete overview of the problems related to the Campania waste system. Using a new system of accounting, i.e. the Multi-Scale Integrated Analysis of Societal and Ecosystem Metabolism (MuSIASEM), they define a novel integrated set of indicators. They focus specifically on the Density of Waste

Disposed (DWD) and show that it can be integrated with the more conventional indicators of waste to provide richer information on Campania's waste crisis. DWD enables a more complete description of waste management systems and the plausible risk of socio-environmental conflicts. Indeed, it helps to explain the biophysical pressure and ecological unsustainability of waste management in the region of Campania. Multi-scale analysis of DWD highlights the diversity in Campania in terms of waste performance by the provinces and shows that regional results are mainly driven by Naples, the capital of Campania, whose DWD is much higher than in Campania's other provinces.

Moreover, the patterns of waste generation and disposal in Campania suggest that rather than imposing the same separated waste collection targets, as is the case across Europe, they should be set according to the biophysical characteristics of different areas.

To explore this further, this chapter addresses the following research question: would the biophysical pressure of waste metabolism of Naples be sustainable if its citizens achieved the target of 65 per cent of separate collection by the end of 2012, as prescribed by the national law?

The chapter is organized as follows: the next section introduces the case study of Campania's waste crisis. The following sections provide an overview of waste generation and separation in Campania and discuss patterns of DWD at different scales; first, Campania's regional pattern is compared with those in other regions of Italy; second, the province of Naples is compared with other Italian provinces; and finally, comparison is made among Campania provinces. In the final part of the chapter we discuss some possible medium-term scenarios related to the achievement of the separated collection target by Campania's provinces and present some conclusions.

Background to the Campania case study

Campania is a region located in the south of Italy and its capital is the city of Naples. The region is comprised of five provinces: Naples, Avellino, Benevento, Caserta and Salerno (see Figure 5.1). The metropolitan area of Naples has the highest density of population in Italy and one of the highest in Europe. Its average population density is around 2,000 inhabitants per km^2; in the central district, which has a population of one million, population density is 8,500 inhabitants per km^2 (ISTAT 2009). The metropolitan area of Naples is a large and complex urban–rural system with over four million inhabitants, where agriculture, food production, industry, and waste treatment and disposal activities coexist.

The waste crisis began officially in 1994. National government declared a state of emergency[1] due to the rapidly decreasing landfill capacity in Campania (the actual number has never been clearly presented and discussed) (D'Alisa and Armiero 2011), and Campania's failure to develop and implement a regional waste plan. A commissioner was appointed and given the power to make fast decisions to cope with the waste management

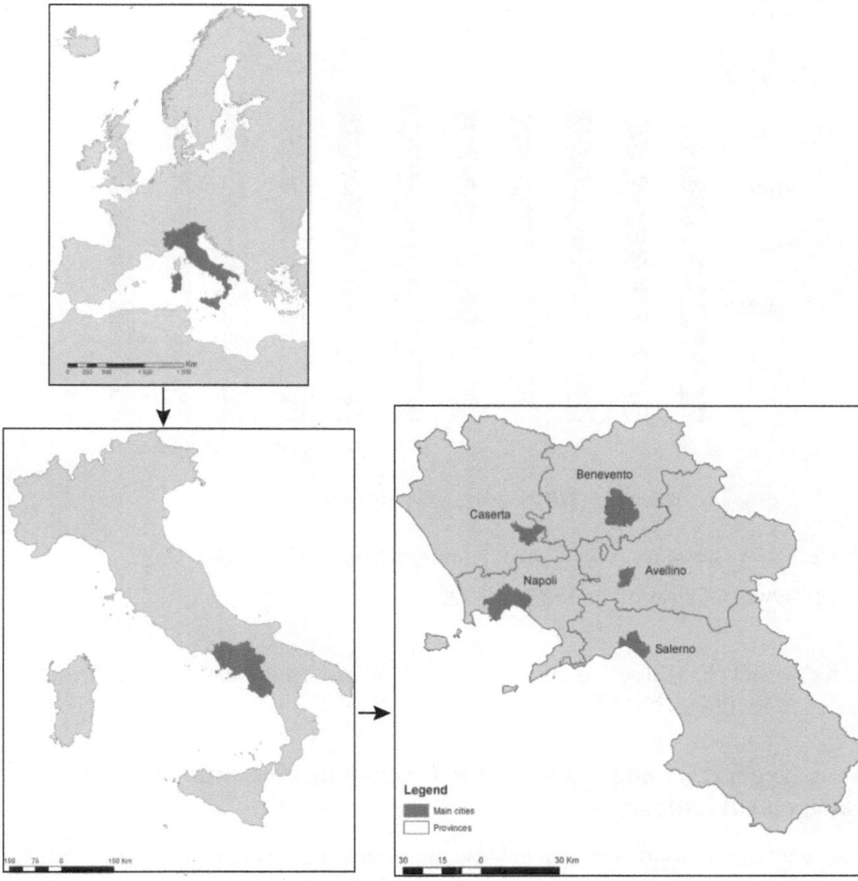

Figure 5.1 Europe, Italy, Campania (maps).
Source: Created by Davide Burgalessi.

problem (for a detailed discussion, see D'Alisa *et al.* 2010). The commissioner set up what was meant initially to be a temporary institution to deal with the extraordinary waste crisis; but it lasted 15 years. This led to the derogation of the general legislative framework and responsibilities, and the entire territory of Campania was forced to implement the same extraordinary measures. The top-down imposition of waste policies, in the name of 'fast' decision-making, resulted in a lack of transparency and participation, and increased social unrest. The effect of this extended 'emergency' has been to crystallize the crisis and make Naples an icon of urban disaster. After 15 years of emergency what was an exceptional situation has become the norm, and the waste mismanagement has been transformed in a crisis of democracy (D'Alisa *et al.* 2010), in which the claims of grass-roots movements have been delegitimized by a supposed relation with the Camorra, the Neapolitan

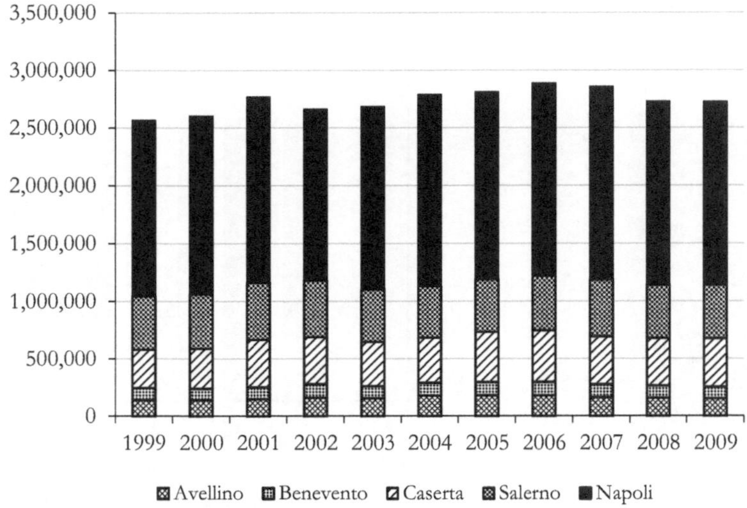

Figure 5.2 Waste generation by Campania's provinces (1999–2009).
Source: Own elaboration on ISPRA data.

mafia, which has blocked the enactment of an effective waste policy (*The Economist* 2008a, 2008b).

Waste generation and separation in Campania: conventional waste management indicators

This section provides an overview of waste generation and separation patterns at the regional and provincial levels. Figure 5.2 shows the evolution of the total waste generated in Campania between 1999 and 2009. Total waste generation has increased from 2.56 million tons a year to 2.72 million tons: by about 12 per cent between 1999 and 2006, and by 6 per cent between 2007 and 2009. This level of increase is lower than the Italian average, which is 14 per cent. Comparison with other Italian regions shows that Campania generates less waste than Lombardy, Lazio and Emilia, where waste growth rates are 15 per cent, 20 per cent and 20 per cent respectively.

The provincial pattern shows high territorial heterogeneity. The regional result is clearly driven by the province of Naples, which is home to 53 per cent of the region's population and is responsible for about 58 per cent of total waste generated. Moreover, between 2002 and 2009, the waste generated in the province of Naples increased about 7 per cent, while in the rest of the province it decreased slightly. A decomposition of the regional data shows that the good performance of some provinces is being counteracted by the high demographic weight of Naples.

Regional waste generated is an absolute value that depends on population size; therefore, it is important to consider waste generation per capita. In 2008, almost 0.47 tons of waste per capita were generated in Campania, an amount lower than the average in Italy and Europe, respectively 0.53 tons, 0.51 tons (EU27) and 0.55 tons (EU15). Waste generation per capita in the province of Naples, 0.51 tons, was lower than in the province of Rome (0.62 tons per capita), Bologna (0.55 tons per capita), Palermo (0.53 tons per capita), and was fairly similar to Milan and Turin (both 0.49 tons per capita) (ISPRA 2009). Among the other Campania provinces, waste generated per capita was lower, for example, Avellino 0.34 tons, Benevento 0.36 tons, Salerno 0.41 tons and Caserta 0.47 tons.

In relation to separated collection, it should be noted that, in 1999, four year after the 'state of emergency' was declared, almost none of the waste generated in the region was separated. From 1 per cent in 1999 and 2000, the regional separation rate increased to 29 per cent in 2009 which was below the national average (34 per cent), but still higher than in Liguria (24 per cent), Abruzzo (24 per cent), Lazio (15 per cent), Puglia (14 per cent), Calabria (12 per cent), Basilicata (11 per cent), Molise (10 per cent) and Sicilia (7 per cent). The province of Naples achieved 24 per cent separated collection, lower than in Avellino and Salerno (48 per cent), Benevento (30 per cent), but higher than Caserta (21 per cent).

This first brief overview of waste performance in Campania highlights several issues:

1. The problem of waste management in Campania is not just a waste generation problem. Regional generation is high, but not the highest in Italy. Growth in waste generation across the time horizon considered is 6 per cent, which is much lower than in the rest of Italy (14 per cent). Waste generation per capita in Campania is also below the national average, respectively 0.47 tons and 0.53 tons in 2009. Thus, residents in Campania generate less waste per capita than the average Italian citizen.
2. The separation rate in 2009 was 29 per cent, below the national average of 34 per cent. However, it is not the lowest in the regional rankings; regions such as Liguria, Abruzzo, Lazio, Puglia, Calabria, Basilicata, Molise and Sicilia achieved lower levels.
3. The regional data hide a territorial diversity, which emerges in the provincial analysis. Waste generation in Naples has increased, while it has slightly decreased in the other provinces, as has waste generation per capita. Also, the separation rate in the province of Naples in 2009 was 24 per cent, lower than in Avellino and Salerno, 48 per cent, which are above the national average, and Benevento, 30 per cent. Unfortunately, the good results achieved in most of the region are hidden by the poor performance of Naples.

These first conclusions highlight that investigating Campania using traditional indicators for waste does not provide explanations for the crisis. Although waste generation is increasing and its absolute value is high, this is consistent with – and sometimes even better than – trends in other regions. Therefore, we need some additional information to conduct an exhaustive analysis of the problem in Campania.

Complementary indicator: density of waste disposed

D'Alisa *et al.* (2012) discuss the limits of conventional indicators and the need to integrate new indicators that are more able to describe waste management systems and the risk of socio-environmental conflict. They propose a new integrated set of indicators based on a new accounting system, that is, the MuSIASEM. In the context of the present study we should mention that DWD refers to the weight of waste disposed[2] in a given area and depends on two factors: demographic pressure and implementation of separated collection – and the effect of both the behaviour and technology used in the process of waste management. It can be calculated as follows:

$$DWD_i = WD_i/S_i$$

where WD is the amount of waste not separated and S is the area under analysis, both at the different hierarchical levels i (regions, provinces and municipalities), or as:

$$DWD_i = WDMR_i * DHA_i$$

The second formulation is within the MuSIASEM framework, according to which WDMR is the Waste Disposed Metabolic Rate[3] and indicates the weight of waste eventually disposed in landfill, either directly or after being burnt in an incinerator. It refers to the environmental loading in terms of required sink capacity (land and air), associated with one hour of human activity. DHA[4] is the Density of Human Activity and represents the demographic density of different human activities (i.e. work, study, mobility, tourism, and leisure and entertainment) at different levels.

DWD can be written also as follows:

$$DWD_i = POP_i/S_i * WG_i/POP_i * WD_i/WG_i$$

This formulation enables comparison between DWD and the IPAT indicator (Ehrlich and Holdren 1971). According to Ehrlich and Holdren, who first proposed this indicator in 1971, environmental Impact (I) is driven by: Population (P), Affluence of the society (A), simplified by consumption per capita, and the Technology (T), defined as the impact produced per unit of production.

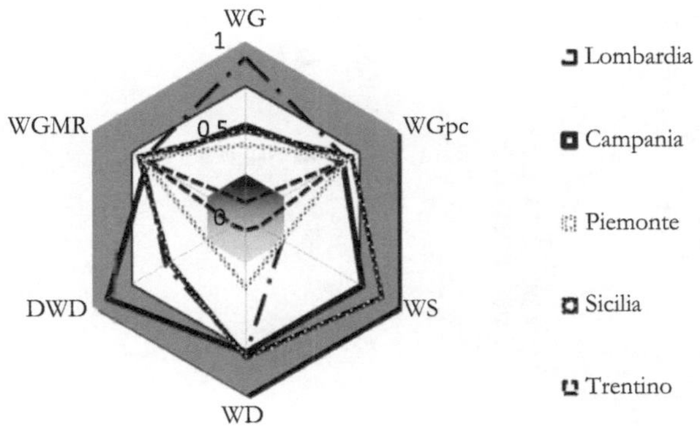

Figure 5.3 Waste indicators by region.

Notes: * The acronyms stand for: WG, Waste Generated; WGpc, Waste Generated per capita; WS, Waste Separated; WD, Waste Disposed; DWD, Density of Waste Disposed; WGMR, Waste Generated Metabolic Rate.

Therefore, the components of DWD can be considered proxies for the components of IPAT. More specifically, population density (POP_i/S_i) is a proxy for population size (P); waste generation per capita (WG_i/POP_i) is a proxy for affluence (A); the fraction of the waste disposed on total waste generated (WD_i/WG_i) is a proxy for the technology (T) implemented.

Analysis based on DWD provides a different picture and richer information than is provided using traditional indicators. D'Alisa *et al.* (2012) present their results by means of a radar chart. The value of each indicator is contextualized by the colour/position of the range in which it falls: grey/centre is good; white is acceptable; grey/contour is unsatisfactory.[5] Figure 5.3 depicts the performance of five Italian regions according to the indicators mentioned so far.[6] It can be seen that Campania does not show any remarkable differences from the other regions based on conventional indicators. The rate of separate collection is lower than in most of the sample, apart from Sicilia. When we look at DWD Campania is the only region in the grey/contour band.

This result is confirmed if the analysis is scaled down to provincial level (Figure 5.4). According to all the indicators, except DWD, the performance of Naples is consistent with the performance of other Italian provinces. Indeed, for density, it shows the worst result and is remarkably distant from the rest of the sample.

The other provinces are white (Milano, Trieste) or grey/centre (Palermo, Torino), which correspond to much lower levels of ecological pressure.

Figure 5.5 provides a comparison of the waste performance among Campania provinces. The change of scale highlights a territorial heterogeneity that is not observable at the regional level. In relation to the indicator for separate collection, the province of Naples lies in the grey/contour zone with Caserta and Benevento, while Avellino and Salerno are in the white

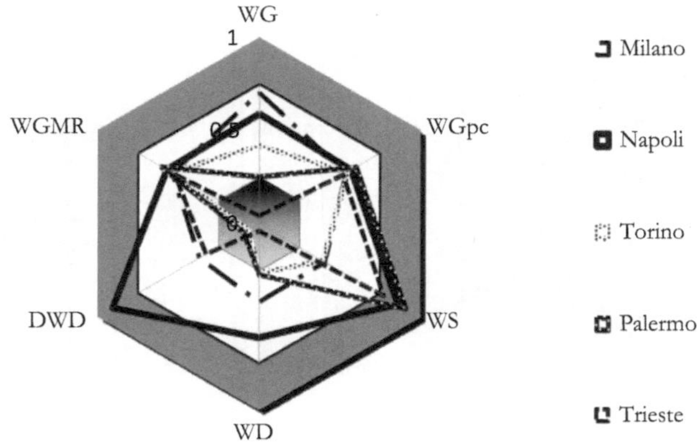

Figure 5.4 Waste indicators by province.

Notes: * The acronyms stand for: WG, Waste Generated; WGpc, Waste Generated per capita; WS, Waste Separated; WD, Waste Disposed; DWD, Density of Waste Disposed; WGMR, Waste Generated Metabolic Rate.

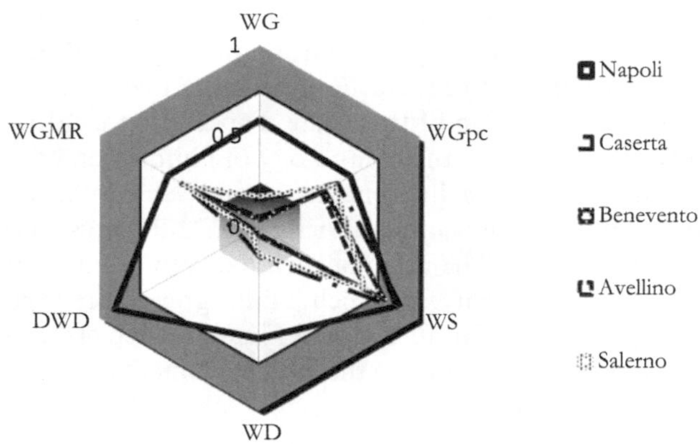

Figure 5.5 Waste indicators by Campania's provinces.

Notes: * The acronyms stand for: WG, Waste Generated; WGpc, Waste Generated per capita; WS, Waste Separated; WD, Waste Disposed; DWD, Density of Waste Disposed; WGMR, Waste Generated Metabolic Rate.

zone. When we consider the indicator DWD, Naples is the only province in the grey/contour zone, the other four are in the grey/centre zone.

The results confirm that the DWD indicator complements and sheds light on the other indicators used to analyse critical waste management patterns; otherwise, looking at the conventional indicators, the waste crisis in Campania is invisible. Waste generation per capita in the region is quite similar to the national average and separation rates are reasonable compared to the other Italian regions. But when we consider DWD we find that Campania has the worst performance in the regional ranking and that the impact of its waste management is very high. Furthermore, if we scale down to the provincial we see that the regional impact is mainly driven by the province of Naples whereas the rest of Campania's provinces show good performance.

Thus, D'Alisa *et al.* (2012) suggest some policy conclusions:

1. The imposition of the state of emergency on the entire region has not been constructive in solving the waste crisis in Campania. Rather than imposing the same emergency plan on all of its provinces regardless of their (good) waste performance, it would have been better to identify the area at risk based on a biophysical indicator rather than on administrative logic.
2. All new building should be banned in the metropolitan area of Naples in order to halt the increase in the already unsustainable high-density population, which is the main driver of DWD levels in the province, and to avoid future socio-environmental conflicts over the localization of waste infrastructures. Alternatively, very restrictive policies on the prevention of waste generation should be imposed to reduce waste per capita by very large amounts.

Of particular interest to the present chapter is that the high value of DWD in the province of Naples corresponds to huge ecological pressure and sheds light on the biophysical roots of the problem underlying the waste generation patterns in the province.

Thus, in the next section we test what would happen if Naples achieved the separated waste collection targets that have been imposed by law.

An overview of the separated collection targets

Before going through the scenario analysis, we briefly discuss the policy targets related to recycling activities in Europe and Italy. One of the main objectives of European waste strategy is that we should become a recycling society. In order to move towards this general goal, the European Commission has prescribed that all member states should take measures to achieve the following separated collection targets by 2020 (Dir. 2008/98/EC):

1. 50 per cent by weight of waste such as paper, metal, plastic and glass;
2. 70 per cent of construction and demolition waste.

Also, according to the landfill directive (Dir. 1999/31/EC as amended 1882/2003/EC and 1137/2008/EC), European countries must reduce the amount of biodegradable waste going to landfill by:

1. 75 per cent by 1996;
2. 50 per cent by 2009;
3. 35 per cent by 2016.

Table 5.1 Targets of separate collection according to different laws and ordinances issued at national and regional level

Laws and ordinances	*Targets of separate collection*
Art. 24, D.Lgs 22/97	15% → 1999 25% → 2001 35% → 2003
Art. 205, D.Lgs 152/06	35% → 2006 45% → 2008 65% → 2012
Art. 1 paragraph 1108 L. 296/06 i.e. the Italian Financial Law 2007	40% → 2007 50% → 2009 60% → 2011
Urban Waste Management Plan Campania 2007	25% → 2007 35% → 2010 50% → 2011–2012
Art. 11 L. 123/2008	25% → 2009 35% → 2010 50% → 2011
Urban Waste Management Plan Campania 2011	50% → 2011

Specific targets have been set for different types of waste, such as: packaging waste (Dir. 94/62/EC as repealed by Dir. 2004/12/EC); end of life vehicles (Dir. 2000/53/EC); waste electrical and electronic equipment (Dir. 2002/96/EC); batteries and accumulators (Dir. 2006/66/EC).

The logic of the European waste legislation was transposed into Italian law by means of legislative Decree 22/1997. This law decreed that every Italian province should reach the 15 per cent target of separated collection by 1999, 25 per cent by 2001 and 35 per cent by 2003 (Article 24, Decree 22/1997). Table 5.1 presents a synthesis of the separated collection targets according to different laws and ordinances issued at national and regional levels, between 1997 and 2011. It can be seen that the targets have been continuously rearranged. For instance, the target of 35 per cent by 2003, according to the law 22/97, was extended to 2006 by the law 152/06. It can also be seen that the target set by the last waste plan proposed by the government of Campania (Regione Campania 2011) is lower than the target imposed by the national law (L. 296/06): respectively 50 per cent and 65 per cent by 2011.

The above overview confirms that the imposition of separated collection targets at EU and Italian level does not take account of the social, economic and environmental characteristics of different territories. The absence of any analytical reasoning behind the setting of targets has allowed politicians, on different administrative levels, to change the amount of separated collection to be achieved.

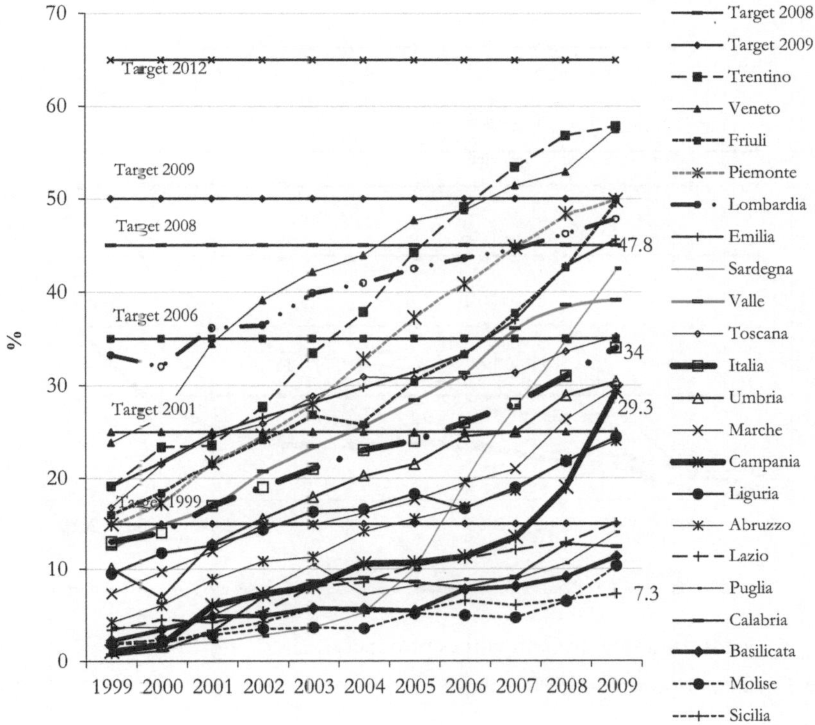

Figure 5.6 Separation rate by Italian region (2011).
Source: Own elaboration on ISPRA data.

Official data show that at the national level these targets have not been achieved. Figure 5.6 illustrates the rate of separated collection achieved by the 20 Italian regions in 1999 to 2009, compared to the targets imposed by different laws and decrees. In 2009 the national rate of separated collection was about 34 per cent. The figure shows the high level of discrepancy among regions, especially between those in the north and the south. Only Trentino and Veneto have separation rates higher than the target established, while Friuli and Piemonte almost achieved it. In 2009 about 80 per cent of the Italian provinces were below the 2009 target and 55 per cent were below the 2006 target. Around 40 per cent of the sample have not achieved the 25 per cent target set for 2001. Sardegna is the only southern region to achieve the target of 50 per cent imposed for 2009. Other southern regions have achieved no more than 30 per cent. Campania's separated collection was 29 per cent in 2009. Other regions have even lower rates, some not even achieving 20 per cent rate and Sicilia achieving less than 10 per cent.

The failure to achieve the separated collection targets applies also within Campania. Figure 5.7 shows the evolution of separation rates in Campania's

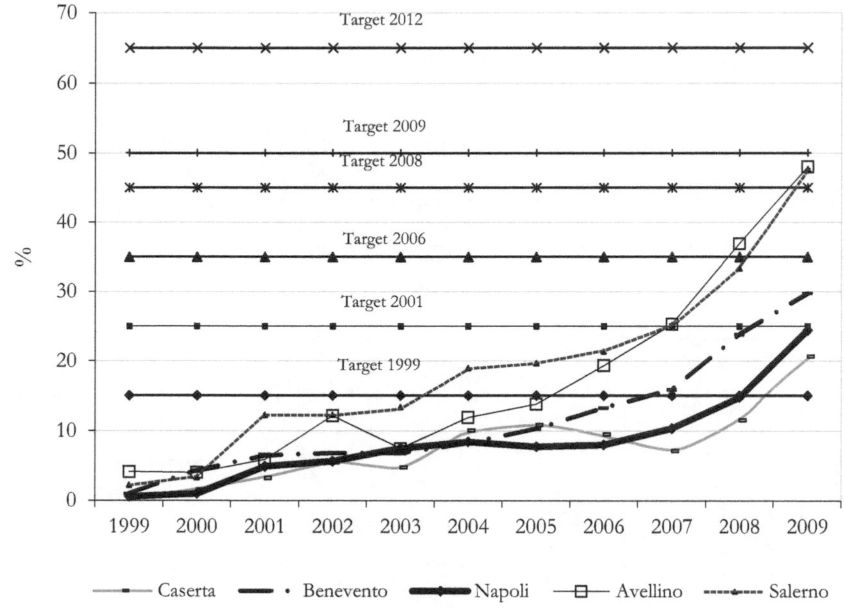

Figure 5.7 Separation rate by Campania's provinces.
Source: Own elaboration on ISPRA data.

provinces compared to the targets imposed by the law (Art. 24, D.Lgs 22/97, Art. 205 D.Lgs 152/06). Up to 2003 none of the five provinces had reached any of the targets imposed. In 2004, all of Campania's provinces except Salerno were lower than 15 per cent (Avellino 14 per cent, Caserta 11 per cent, Benevento 10 per cent, Naples 8 per cent). There was an important change to this patter in 2007–8 when Avellino and Salerno achieved respectively 37 per cent and 33 per cent, close to the 2006 target of 35 per cent, but a long way from the 2008 target. In 2008, Naples and Caserta were achieving lower rates than the targets imposed for 1999. In 2009 Avellino and Salerno were close to 50 per cent of the target imposed by the law, while Benevento was still below the target set for 2006. Naples and Caserta, despite better performance compared to 2008, are still below the 2001 target. It should be noted also that the ecological pressures and levels of (un)sustainability are very different in these provinces. In 2007, for example, DWD for the province of Naples was 3,462 kg/d/km^2, while in Avellino, Benevento, Salerno and Caserta the levels were respectively 115 kg/d/km^2, 123 kg/d/km^2, 201 kg/d/km^2 and 401 kg/d/km^2,

Scenario analysis

We next analyse different medium-term policy scenarios. The software package Vensim (1988) is used to perform the simulations.

Table 5.2 Separation rate growth (to reach the 65 per cent target by 2012)

Province	*Separation growth rate*
Naples	39%
Avellino	11%
Benevento	30%
Caserta	45%
Salerno	11%

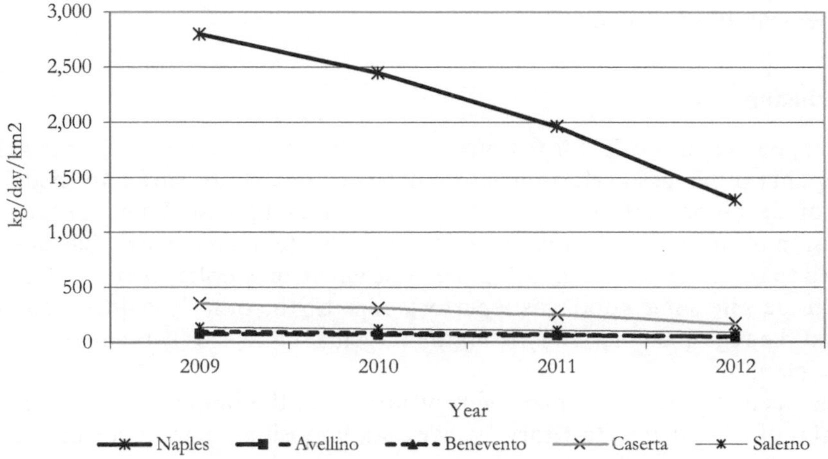

Figure 5.8 Evolution of DWD in Campania.

The first simulation analyses the growth in separated collection. The first simple test aims to identify the level of yearly separation growth rate to be achieved by each of Campania's provinces if the 65 per cent target is to be achieved by 2012, as established by the national law (Art. 205, D.Lgs 152/06). The results are presented in Table 5.2 and show that the province of Naples would need to increase its separated collection rate by 39 per cent per year, the province of Caserta by 45 per cent, Benevento by 30 per cent, and Salerno and Avellino by 11 per cent per year.

The rates provided by the simulation are used to analyse the evolution of DWD within Campania from 2009 to 2012. Figure 5.8 shows that if separated collection increased at 39 per cent per year, DWD for Naples would decrease to 1,295 kg/day/km^2 in 2012. Although the value is 54 per cent lower than in 2009, it will be slightly higher than the value of the second worst province of Italy in 2007, which is Milan (1,406 in 2007), and is still higher than all the other bad performers at provincial level such as Trieste (1,250 in 2007) and Rome (1,126 in 2007). We observe also that DWD in Naples in 2012 would be much higher than the 2009 starting value for the rest of provinces. Figure 5.8 shows that the density for Caserta would drop from 349 kg/day/km^2 to 163, Salerno from 134 to 90 kg/day/km^2, Benevento from 96 to 48 and Avellino from 76 to 51 kg/day/km^2.

This simple analysis shows that if Naples were to achieve the target of 65 per cent of separated collection imposed by the law, its DWD levels would still be very high compared to the rest of Italy and the other Campania provinces.

Finally, we compare density for Naples with the Italian average (Figure 5.9). The results of the simulation show that, if separated collection increases at 39 per cent, DWD in Naples would approach the average DWD in Italy, between 2014 and 2015, but this would require an impossible level of almost 100 per cent of separated collection!

Conclusion

The huge value of DWD for the province of Naples compared to the other Campania provinces corresponds to a huge ecological pressure and sheds light on the biophysical roots of the waste crisis in that region. It implies that inclusion of the entire Campania region in the 'state of emergency' has not helped to solve the crisis. Imposing the same emergency policy on the whole region, despite some good waste performance by the other provinces, has moved the region farther away from the targets established by the regional waste plan.

The need to tune waste management targets to the biophysical characteristics of a particular territory, rather than imposing a purely normative stance, was proposed in this work with respect to separated collection policy. EU policy so far has imposed the same separated collection targets on all member states. It has taken no account of the different social, economic and ecological characteristics of each region. It underestimates the fact that regions with different societal characteristics produce different environmental loading. The Campania case study shows that if Europe wants to move towards a sustainable pattern its environmental policy should be tuned also to biophysical indicators, to construct an environmental friendly society. The different territorial benchmarks of (un)sustainability should articulate, at different levels, more diversified separated collection targets to achieve different waste management activity organization. We suggest that one of these benchmarks might be DWD. It implies that the higher the value of DWD the higher should be the separated collection target and the stricter the waste prevention policies. In order to demonstrate that a simple normative target, not based on biophysical analysis, makes no sense from an environmental point of view, we proposed a scenario in which the province of Naples achieved the targets of 65 per cent of separated waste collection imposed by the law. The results show that the environmental load of the region would still be very high compared to the remaining Italian and Campania provinces. DWD would be the highest in Italy at the

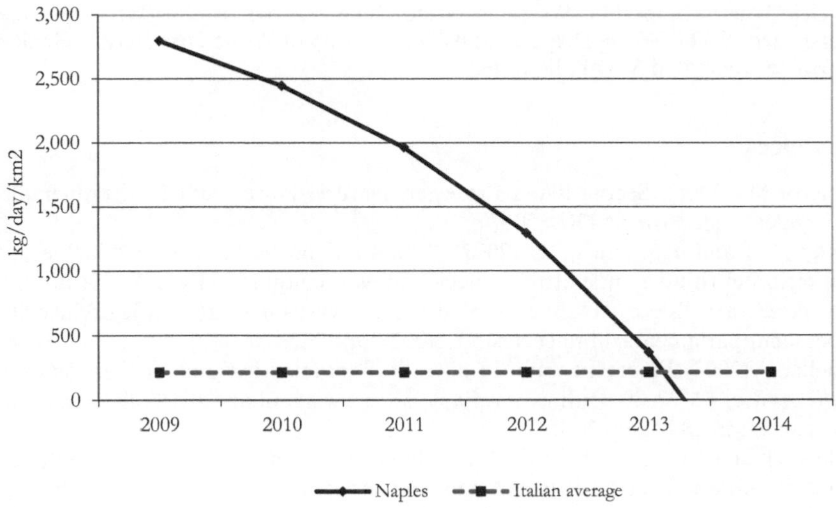

Figure 5.9 Evolution of DWD in Naples compared to the national average.

provincial level. This means that 65 per cent of separated waste collection, were it possible for the province of Naples, would not be sustainable. At the same time, the ecological burdens (DWD) of other provinces in Campania, such as Benevento and Avellino, are among the lowest in Italy, despite not achieving the targets. This questions why different territories characterized by different societal, economic and ecological patterns should have the same environmental targets. We need to move forward and re-articulate environmental performance on the basis of the different burdens produced by their different socio-economic metabolisms. This could be achieved by using a biophysical benchmark. Then, future European and national policies related to separated waste collection, rather than imposing the same target on all countries regardless of their territorial heterogeneity, should be fine-tuned to the different biophysical characteristic of their administrative parts and a multi-separated waste collection target should be established as our analysis of the waste metabolism in Campania quite clearly shows.

Notes

1 In this work, the term 'emergency' implies 'extraordinary', beyond ordinary institutional measures. We use this term following Mastellone *et al.* (2009) and D'Alisa *et al.* (2010).
2 In their work, Waste Disposed is the amount of waste not separately collected.
3 WDMRi = WDi/ HAi where WD is the amount of non-separated waste and HA is human activity.
4 DHA_i is Density of Human Activity and is a proxy of the demographic density at the different levels *i*.
5 For further details on radar charts see D'Alisa *et al.* (2012).

6 WG is Waste Generated; WGpc is Waste Generated per capita; WS is Waste Separated; WD is Waste Disposed; DWD is Density of Waste Disposed; WGMR is Waste Generated Metabolic Rate.

References

Armiero, M. (2008) 'Seeing like a Protester: Nature, Power, and Environmental Struggles', *Left History*, 13(1): 59–76.

Barbieri, R. and Piglionica, D. (2007) 'Commissione parlamentare d'inchiesta sul ciclo dei rifiuti e sulle attività illecite ad esso connesse' (Doc. XXIII no 2). Parliamentary Report. Online. Available at: www.camera.it/_dati/leg15/lavori/documentiparlamentari/indiceetesti/023/002_S/pdfel.htm.

Comella, G. (2007) 'Emergenza rifiuti e mortalità per tumori in Campania', *Bollettino delle Assise*, 12: 27–39. Online. Available at: www.napoliassise.it/bollettino.htm (last accessed 28 March 2010).

D'Alisa, G. and Armiero, M. (2011) 'La ciudad de los residuos. Justicia ambiental e incertitumbre en la crisis de los residuos en Campania (Italia)', *Ecología Política*, 41: 97–105.

D'Alisa, G., Di Nola, M.F. and Giampietro, M. (2012) 'A Multi-Scale Analysis of Urban Waste Metabolism: Density of Waste Disposed in Campania', *Journal of Cleaner Production*, 35: 59–70.

D'Alisa, G., Burgalassi, D., Healy, H. and Walter, M. (2010) 'Conflict in Campania: Waste Emergency or Crisis of Democracy', *Ecological Economics*, 70: 239–49.

de Medici, G. (2007) 'L'emergenza rifiuti in Campania la questione dei siti', *Bollettino delle Assise*, 12: 6–7. Online. Available at: www.napoliassise.it/bollettino.htm (last accessed 28 March 2010).

The Economist (2008a) 'Rubbish in Naples. See it and Die. The Real Crisis in Naples is about Governance as Much as Rubbish', 10 January 2008. Online. Available at: www.economist.com/node/10499135.

The Economist (2008b) 'The Mafia in Naples. Gangsters go Global', 10 January 2008. Online. Available at: www.economist.com/node/10493264.

Ehrlich, P.R. and Holdren, J.P. (1971) 'Impact of Population Growth', *Science*, 171: 1212–17.

Fazzo, L., Belli, S., Mitis, F., Santero, M., Martina, L., Pizzuti, R., Comba, P. and Martuzzi, M. (2008) 'Analisi dei clusters di mortalità in un'area con una diffusa presenza di siti di smaltimento dei rifiuti urbani e pericolosi in Campania', Istituto Superiore di Sanità. Online. Available at: www.iss.it/binary/epam/cont/FAZZO_Rifiuti.1159881860.pdf.

Fontana, E., Pergolizzi, A., Ruggiero, P., Dodaro, F., Groccia, C., Ciafani, S. and Del Giudice, R. (2008) 'Rifiuti Spa. Legambiente'. Online. Available at: www.borsarifiuti.com/materiali.phpsc?i=d (last accessed 28 March 2010).

Iacuelli, A. (2007) *Le vie infinite dei rifiuti. Il sistema campano*, Italy: Edizioni Rinascita.

ISPRA, Istituto Superiore per la Protezione e la Ricerca Ambientale (2001–2011) *Rapporto Rifiuti*. Online. Available at: www.apat.gov.it/site/it-IT/APAT/Pubblicazioni/Rapporto_Rifiuti.

ISTAT (2009) *Statistiche Demografiche*. Online. Available at: http://demo.istat.it/.

Lucarelli, A. (2007a) 'Riflessioni sul piano rifiuti della Campania', *Bollettino delle Assise*, 12: 8–17. Online. Available at: www.napoliassise.it/bollettino.htm.

Lucarelli, A. (2007b) 'Governare e gestire la raccolta differenziata', *Bollettino delle Assise*, 6, 7, 8 and 9: 21–30. Online. Available at: www.napoliassise.it/bollettino.htm.

Lucarelli, A. (2007c) 'Profili di illeggittimità ed inopportunità del piano regionale dei rifiuti della Campania', *Bollettino delle Assise*, 6, 7, 8 and 9: 5–20. Online. Available at: www.napoliassise.it/bollettino.htm.

Martuzzi, M., Bianchi, F., Comba, P., Fazzo, L., Minichilli, F. and Mitis, F. (2008) 'Trattamento dei rifiuti in Campania. Studio di correlazione tra rischio ambientale da rifiuti, mortalità e malformazioni congenite'. Online. Available at: www.iss.it/binary/epam/cont/view.1190959279.htm.

Mastellone, M.L., Brunner, P.H. and Arena U. (2009) 'Scenarios of Waste Management for a Waste Emergency Area: a Substance Flow Analysis', *Journal of Industrial Ecology*, 13: 735–57.

Musella, A. (2008) *Mi rifiuto! Le lotte in difesa della salute e dell'ambiente in Campania*. Dogliani: Edizioni Sensibili alle Foglie.

Nicolli, F., Mazzanti, M. and D'Amato, A. (2011) 'Waste Sustainability, Environmental Policy and Mafia Rents: Analysing Geographical and Economic Dimensions', paper presented at 9th ESEE conference, Istanbul. Online. Available at: www.esee2011.org/program/print.sessions.v2.php.

Ortolani, F. (2008) 'Emergenza rifiuti in Campania: due discariche a rischio frana, la terza inquina: Ferrandelle non era idonea'. Online. Available at: www.9online.it/.

Rabitti, P. (2008) *Ecoballe. Tutte le verità su discariche, inceneritori, smaltimento abusivo dei rifiuti*, Roma: Aliberti editore.

Raimondi, R. (2007) 'Rapporto sul disastro ambientale dei rifiuti in Campania. Il diritto al risarcimento dei danni', *Bollettino delle Assise*, 12: 29–33. Online. Available at: www.napoliassise.it/bollettino.htm (last accessed 28 March 2010).

Regione Campania (2011) 'Sintesi della Proposta di Piano Regionale per la Gestione dei Rifiuti Urbani della Regione Campania', March 2011. Online. Available at: http://orr.regione.campania.it/osservatorio/docs/documenti/Doc_1081.pdf (last updated November 2011).

Senior, K. and Mazza, A. (2004) 'Italian "Triangle of Death" Linked to Waste Crisis', *The Lancet Oncology*, 5: 525–27.

Vensim® (1998) *3.0 Reference Manual, Ventana Systems*, Vensim®.

Part II

The international setting

Waste trade drivers, convergence and policy making in spatial-framed environments

6 International waste trade

Impacts and drivers

Massimiliano Mazzanti and Roberto Zoboli

Introduction

The international trade in waste is increasing. In the case of the EU, the export of 'notified waste' increased from 6.3 million tons in 2001 to 11.4 million tons in 2009 (Figure 6.1).[1] A large share of the total flows is represented by intra-EU flows (both notified and non-notified waste) although extra-EU flows are increasing.

Although the amount of international waste trade (IWT) is a small part of total waste production, its rising trend raises two logically different, although dynamically interconnected, questions relevant to research and to policy:

1. What are the environmental and economic impacts of IWT?
2. What are the drivers of IWT?

The first question reflects the many concerns raised by debates on the international circulation of waste. There are *economic* and *environmental impacts* associated with IWT. These impacts, in particular those on the environment, can have direct implications for the legal and policy regime for IWT in Europe (and worldwide), which focuses mainly on minimizing environmental impacts. The environmental and economic impacts of actual IWT, that is, from an *ex post* perspective, can be addressed independently from decisions to trade waste internationally rather than manage it domestically, although the two may be dynamically linked.

Understanding IWT *drivers* requires careful analysis of the economic and institutional factors, including different countries' environmental regulation, that explain why and how IWT occurs. This analysis can have important policy implications: by highlighting the decision mechanisms related to IWT we can suggest how European and national policies can influence IWT in the desired direction.

This chapter proposes a framework for the analysis of IWT at both levels – impacts and drivers – to address the questions set out above. The availability of formal models of environmental and economic impacts and

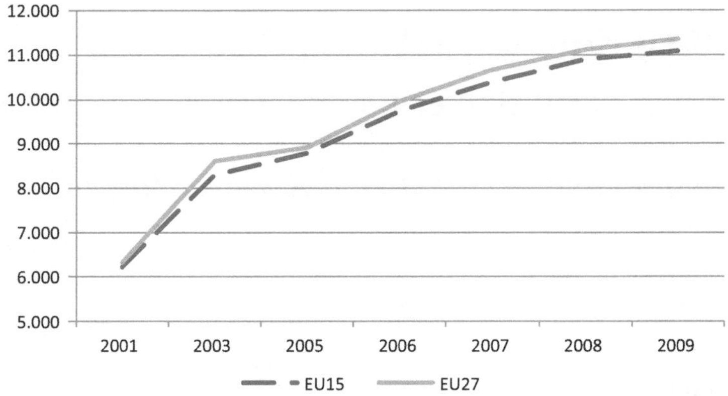

Figure 6.1 Export of notified waste (hazardous and non-hazardous) from EU27 countries, including intra-EU trade, 2001–2009 (000 tons).

Source: Eurostat.

drivers of IWT is limited. We propose a framework for analysis and discuss some issues related to both methods and applications. In particular we propose: (1) a logically consistent scheme of accounting and measuring of the net environmental impacts arising from IWT; the scheme is based on the idea of 'differential impact', which can be applicable to all international contexts; (2) a scheme for accounting the value added gains and losses associated with IWT compared to domestic management – also based on a differential approach; (3) a set of drivers that may explain why IWT does arise. The development of empirical analyses based on this framework requires some information that is not presently available; thus, we do not provide model-based applications. However, we suggest an example of a specific IWT flow that highlights how the main drivers of trade do work.

Environmental impacts: a 'differential approach'

We provide the outline of an impact evaluation scheme for:

- a given quantity (one ton) of a well defined category of homogeneous waste collected in the home country;
- a specific management/treatment option, for example, landfill, in both home and destination countries (i.e. no change of management option occurs with trade);
- a single category of impact, for example, greenhouse gas (GHG) emissions;
- direct impact only of the first treatment in either the home or destination country;

- a specific single transportation mode, for example, road, which generate impacts in proportion to the distance travelled;
- the assumption of full information, that is, the geography of waste production/collection and management (facilities) across countries is known, allowing transportation distances to be computed.

There are two main broad sources of impacts associated with this IWT: (1) management; (2) transport. The relevant management and transportation impacts are those arising specifically from international trade *only*. This requires a 'differential approach' based on the comparison between the impact from international trade and the impact of non-trade, that is, domestic management, of the *same waste flow*.

We are interested in two different levels of impact: (i) overall (net) impact across countries, for example, the net impact of a flow from Italy to Germany; (ii) the distribution of impacts for each country involved, for example, Italy, Germany and transit countries, if any.

Net impact

The application of a 'differential approach' to the cross-country net impact requires comparison between: (a) the environmental conditions of management/treatment inside the borders of the home/shipping country and management/treatment in the destination country, and (b) comparison of transportation distance possibly travelled by waste if managed inside the home country and distance actually travelled to be managed in the destination country.

Impacts in management. Waste management operations (in our case just one, see assumptions) generate impacts wherever performed. Therefore, assuming trade flow takes place, the *likely* management impact in the country of origin in the case that the waste stream had been managed domestically, must be compared to the *actual* management impact in the place of destination/management abroad. In fact, depending on technologies and environmental rules prevailing in the home country and destination country, international trade may either reduce or increase overall environmental impacts at the management stage. In short, for each category of impact:

$$NMI = IMD - IMH \quad (1)$$

where:

NMI = net environmental impact of IWT in management;
IMD = impact of management at destination country;
IMH = impact of equivalent management in home country.

While the conditions in the country of destination can be measured, those in the home country are necessarily hypothetical, that is, a counterfactual about those at the site/facility where the management *might* hypothetically have taken place domestically (but did not). To obtain this counterfactual reference, we can assume there is always management capacity within the home country – whatever the distance of the facility from the source of waste flow, which is accounted for in the transportation impact (see below). Actually, total lack of capacity or complete ban on disposal/treatment in the country of origin (for the particular waste stream) applies to a limited number of cases. Even where this is the case, the counterfactual can be taken as the possibility of illegal disposal in the home country with associated impacts.

Transportation impacts. A 'differential approach' requires estimation of the net additional transportation distance travelled because of international trade *only*. The distance between the site of production/collection and the site of possible management in the home country must be compared with the distance between the site of production/collection and the site of actual management in the destination country. Net environmental impacts (emissions, energy, etc.) are related to this distance.[2] This differential approach for transport, given the type of treatment and its impacts, is important because it might be that transportation impacts are *reduced* by IWT. Suppose there is a collector in a border region in Italian Alps that ships to Austria for disposal; if the nearest facility in Italy for the same disposal operation is in Sicily (>1,000 km), there can be substantial saving on transportation impact, taking as given the different impact of treatment technologies at the two sites (accounted for in point (a) above). In short:

$$NTI = ITD - ITH \quad (2)$$

NTI = net impact of IWT in transport;
ITD = impact of transport at destination abroad;
ITH = impact of hypothetical transport in home country.

Similar to management conditions, the distance possibly travelled in the home country is necessarily hypothetical, that is, the distance to the home country site/facility where management might have taken place (but did not).

Total impact. The net total impact (NI) of IWT only (for a specific impact category, for example, GHG) is the combination of equations (1) and (2), or:

$$NI = NMI + NTI = (IMD - IMH) + (ITD - ITH) = (IMD + ITD) - (IMH + ITH) \quad (3)$$

The two sources of impact are clearly distinguished and added (same quantity and category of waste, same category of impact), and can be analysed both separately and together. Of course, in the case that the additional transportation distance for international trade is small or even negative (which cannot be ruled out), and the technology of management in the destination country is better than in the home country, IWT might imply positive net impacts on the environment.

The estimate can be replicated for:

- each different category of impacts for the same waste category, which, in general, cannot be summed unless they can be reduced to indexes, for example, toxicity;
- for different management options, for example, recycling or disposal, for the same waste flow;
- for each different category of waste, for example, MSW or hazardous;
- for different transportation modes and their combinations, for example, rail and/or road.[3]

The scheme can be adapted to management options in the destination country, for example, energy recovery, different from that in the home country, for example, disposal in landfill. This scope of application can be best managed if we consider only the direct impact of first treatment in the home and destination country, that is, excluding the impacts along different full disposal/recovery/recycling chains.[4]

Distribution of environmental impacts across countries

The same approach can be used to analyse the impacts in each single country involved in the trade. Assume there are two countries involved: home country or H, for example, Italy, and destination country D, for example, Austria (no transit countries). Assume that the impact of transportation can be allocated in proportion to the mileage of transport within the country borders, that is, there is not transboundary transfer of pollutants via transport.

In country H, export to country D implies two effects (compared to domestic treatment): (i) saving on the impacts of management at home, that is, – IMH; (ii) impacts of net additional transportation distance from the hypothetical site of management at home and transportation to the national border, that is ITB – ITH, where ITB is the impact of transport until border; this difference may even be negative.

Then, in the home country, net impact NH will be:

$$NH = -IMH + (ITB - ITH) \qquad (4)$$

In the country of destination D, the import of waste implies two effects (compared to non-import): (i) the impact of managing waste in the country, that is, + IMD; (ii) the impact of transportation inside the border of the country, or ITBD.

Then in the destination country, net additional impact (ND) is:

$$ND = IMD + ITBD \qquad (5)$$

The exporting country unambiguously gains environmental impacts in management whereas the net addition of the transportation impact can be ambiguous: very high if management was available at a short distance inside the country; low or even negative if the domestic management site was very far from the site of waste production/collection. In average conditions, we can expect that the home country will gain in terms of environmental impacts. The destination country instead loses from the additional impact of waste management and loses from the additional transportation impacts inside the country; neither would have occurred without international trade. Therefore, the destination country unambiguously loses.

Of course, the combination of gains/losses in the two countries corresponds to the net effect across countries defined in Equation (3):

$$NH + ND = (IMD + ITBD) - IMH + (ITB - ITH) = (IMD - IMH) + (ITD - ITH) = NI \quad (6)$$

where ITD = ITB + ITBD, or total transport impact is the summation of impact inside both home country and destination country.[5]

If there are more than two countries involved, for example, a flow from Italy to Germany via Austria, with no management/treatment in the transit country, then the transit country suffers only transportation impacts and risks in proportion to the distance travelled within its borders. In the above equations, total transportation distance TD and the associated impact ITD can be reformulated into:

$$\text{ITD} = \text{ITB} + \text{ITT} + \text{ITBD} \quad (7)$$

where ITB = impact transport inside home country border; ITT = impact of transport in transit countries; ITBD = impact of transport inside border of the destination country.

The net overall impacts within a cross-country perspective remain the same as defined in Equation (3).

Economic impacts: value added creation

A similar 'differential approach' can be adopted to define the *ex post* net economic benefits of IWT, if any. In investigating the net economic-system benefits, we can address the net value added (VA) from market transactions that is associated to *trade only* (i.e. destination country management versus home country management). As in the case of environmental impacts, this analysis can be applied to the net VA creation for the two countries together, and to the distribution of VA between them.

Note that we are considering *ex post* VA creation for the two economies and not the net advantages of single industrial actors (waste collectors/producers) in exporting waste rather than treating it in the home country; these microeconomic variables are addressed in the discussion on drivers in the next section.

Net value added

To simplify the problem, we make the following assumptions. First, the boundaries of VA analysis are three economic actors or phases in the value chain: (i) the home country collector, that is the actor having the ownership of the collected waste; (ii) the treatment facility, located in the home country or abroad; (iii) the transportation company, which will the same for transportation in the home country or abroad; for each of these actors, the VA generated is the difference between the price of their output and the cost of inputs (excluding, of course, labour and capital, the compensation of which is part of the VA); the net value added in both the domestic treatment option and export for treatment abroad is the sum of the three VA components, then net VA at the system level will be the 'consolidated account' (i.e. net of inter-industry exchanges.

Second, we address the VA of the first treatment operation only (at home or in the destination country) and not the possible subsequent stages of industrial transformation of waste.

Third, prices of the waste-derived material (after treatment), costs of waste inputs, and costs of other inputs are all defined per unit of waste and they do not depend on the quantity of waste treated.

Fourth, waste is the only economic input to the treatment facility at home and abroad; its cost (i.e. the price at which waste input is procured by the treatment facility) can be positive or negative depending on the type of waste and its market; therefore, the collector of waste (in the home country) can deliver the waste at either positive or negative prices to the treatment facility (either at home or abroad).

Fifth, there is no capacity constraint on waste treatment in either the home or the foreign country.

Sixth, the price paid for waste inputs by treatment facilities is always the *domestic* price, whether the waste originates in the domestic country or abroad; this can create a different waste price for the home country collector in the two different locations of treatment.

Seventh, the material produced from a unit of waste by the treatment facility is additional and does not substitute for already produced material; similarly, the unit of waste input is additional and does not substitute for another unit of waste in either the domestic or foreign treatment facility.

Eighth, transportation of waste creates economic value added for the transportation industry; transportation cost are supported by the actor (the collector) that ships waste at a treatment facility either at home country or abroad; then, the treatment facility does not have transportation costs of waste procurement as an input; the mode of transportation is the same for delivery at home and abroad, which means that total transportation costs and value added for a unit of waste depend only on the distance travelled.

Within these assumptions, the net value added (VA) creation in the case of *domestic treatment* of the unit of waste is the sum of the three VA components in the three stages (collection, transport, treatment), that is:

$$NVAH = (PH - PWH) + (PWH - ICH - TCH) + (TCH - ICTH) \quad (8)$$

where:

NVAH = net value added for treatment in the home country;
PH = price of the material/energy from waste in the home country;
PWH = price of the waste input in the home country;
ICH = intermediate input cost for collection in the home country;
TCH = transportation cost for delivery to the home treatment facility;
ICTH = intermediate input costs of transport services in the home market;
PH – PWH = VA on treatment at home;
PWH – ICH – TCH = VA on collection at home;
TCH – ICTH = VA on transportation for domestic delivery.[6]

The net VA of the domestic option (NVAH) is the 'vertically consolidated' account resulting from the fact that the price of waste input is a cost (-VA) for the treatment facility while it is a revenue (+VA) for the collectors; similarly the cost of transportation services is a revenue (+VA) for the transportation company while it is a cost for the home country collector (-VA). Then:

$$NVAH = PH - ICH - ICTH \quad (9)$$

The net VA from the domestic treatment option is the difference between the price of the waste-derived material at home (after treatment) and the cost of the intermediate inputs for collection and transportation in the home country.

For the *same waste unit* collected in the home country, net VA in the case of export for treatment in the destination country will be:

$$NVAD = (PD - PWD) + (PWD - ICH - TCD) + (TCD - ICTH) \quad (10)$$

where:

NVAD = net value added for treatment in the destination country;
PD = price of the material/energy from waste in the destination country;
PWD = price of the waste input in the destination country;
ICH = intermediate input cost for collection in the *home* country, origin of the waste;
TCD = transportation cost for delivery to the foreign treatment facility;
ICTH = intermediate input costs of transport services in the home country;

PD – PWD = VA on treatment at a destination in a foreign country;

PWD – ICH – TCD = VA on collection in the home country for delivery of the waste abroad;

TCD – ICTH = VA on transportation for delivery to the destination country.

The different variables in Equation (10) (export) compared to (8) (domestic treatment) stem from the assumptions that: (i) the waste-derived material will be sold in the destination country at PD (i.e. no re-export); (ii) the treatment facility in the destination country will pay the domestic price (PWD) for imported waste inputs; (iii) the home country collector will have the domestic cost of collection (ICH) in both domestic treatment and export options; (iv) the home country collector will bear the total costs of transportation to a foreign destination, which will differ from the cost of transportation domestically only if the distances are different.

The vertically consolidated account of VA for the option of export for treatment is:

$$NVAD = PD - ICH - ICTH \tag{11}$$

The net VA from the 'export for treatment' option is the difference between the price of the waste-derived material in the *destination country* (after treatment) and the cost of the intermediate inputs for collection and transportation by the *home country* collector.

To define the possible net gain on consolidated VA across the two countries (NNVA), we need to look at the difference between the NVAD for treatment of the waste in the destination country and the NVAH from domestic treatment:

$$NNVA = NVAD - NVAH = (PD - ICH - ICTH) - (PH - ICH - ICTH) = PD - PH \tag{12}$$

A net creation of VA from trade will occur if the price of the waste-derived material after treatment in the destination country is higher than the price of the same material in the home country where the waste originated, or:

$$NNVA > 0 \text{ if } PD > PH$$

This results depends from the 'vertically consolidated' features of VA accounting and from the fact that we are adopting a 'differential approach' in which the same waste flow collected in the home country – by the same collector – and transported domestically or abroad – by the same transportation company – has two options for treatment, i.e. domestic facility or foreign facility. Therefore, the *ex post* net VA gain from trading for the two economies combined compared to non-trading, does not depend on the inter-country or inter-facility differences in waste input prices and different transportation costs because both are the costs of intermediate inputs from a value chain perspective.[7] Therefore, while the prices paid for waste as an input in facilities located in different countries can be very important for the *ex ante decisions to trade* and the same importance can be attached to transportation costs (see next section on drivers), they are not relevant for measuring the net economic value creation from waste trading in an *ex post* 'consolidated' accounting perspective of VA.

Value added distribution between countries

The consolidated accounting of net VA can hide significant differences in the distribution of VA from trade compared to domestic treatment. Using the same symbols as in the previous paragraph, and recalling that we are interested in the effects of trade only (differential approach), trading for treatment will create the following effects on VA in the *home country* compared to domestic treatment:

$$NVATradeH = -(PH - PWH) + (PWD - PWH) \quad (13)$$

In the case of export for treatment, the home country will unambiguously lose the treatment VA (PH – PWH) but will possibly enjoy additional VA from exporting the waste if the price for waste as an input in the country of destination (PWD) is higher than the price of waste in the domestic treatment industry (PWH).[8] If we simplify Equation (13), we can see that there will be a gain in VA for the home country in the export for treatment option if:

$$NVATradeH > 0 \text{ if } PWD > PH$$

that is, if the price received for the raw waste exported for treatment abroad is higher than the price of the waste-derived material in the home market. This possibility seems unlikely under normal circumstances since it would mean that, for example, the aluminum derived from one ton of aluminum scrap in the home country is worth less than one ton of raw aluminum scrap exported. Although we cannot exclude this possibility given the huge differences between the materials industries across EU countries, it is more likely that the home country economy will lose VA from trading for treatment abroad – which does not mean that the collector in the home country will lose from exporting rather than delivering the waste to a domestic facility (see the example in footnote seven). It must also be noted that for certain waste flows the price is negative (i.e. the owner has to pay a gate fee to deliver the waste for treatment) and this makes the possibility of gaining from waste trade even more unlikely.

In the case of the destination country, the option to trade the unit of waste will generate the following additional VA:

$$NVATradeD = PD - PWD \quad (14)$$

The destination country will *unambiguously gain VA from importing the waste*, the gain being the difference between the domestic price for waste-derived material and the price of the waste input coming from abroad, under the assumption that this waste input is additional and not a substitute for a unit sourced domestically in the destination country. The gain in VA to the destination country will go entirely to the treatment stage (assuming that the cost and VA of transportation service is allocated to the home country).

Of course, if we add together Equations (13) and (14) we get the same result as in Equation (12), that is, the net VA gain possibly associated with trade for the *two countries together* is the difference between PD and PH:

$$\begin{aligned} NNVA &= [PD - PWD] + [-(PH - PWH) + (PWD - PWH)] \\ &= PD - PH \end{aligned} \tag{15}$$

Merging environmental and economic impacts

In principle, in the case that the environmental impacts examined earlier could be translated in monetary value, it might be possible to combine the net environmental gain/loss and the net VA gain/loss to assess an 'integrated social value change' from IWT, compared to non-trading in a differential approach. This integrated analysis can be done for the two countries together and for each separately. This exercise requires that the monetary value of environmental losses is estimated as VA equivalent, which is not always the case, as suggested by the few EU-scale analyses of external costs (EEA 2011) This exercise of integrated evaluation is not developed here; we just comment on the results in the earlier sections.

We concluded earlier that, under our assumptions, it is not possible unambiguously to claim in advance that IWT involves a net additional environmental impact for the *two countries together* compared to the management of waste at a home country facility. In particular, in the case that the *additional* transportation distance for international shipment is small or even negative (which cannot be ruled out), and the management technology in the destination country is better than that in the home country, IWT might imply positive net impacts on the environment. We then concluded that, under our assumptions, IWT can create net economic VA for the two countries *together* if the price of waste-derived material (or energy) is higher in the destination country than in the home country, where the waste originated. This is not always the case and we cannot say in advance if net VA will be created by trade for the two countries together.

The conclusions about the distribution of environmental and economic impacts between the two countries are slightly less ambiguous. In the case of environmental impacts, we noted that the shipping or home country unambiguously gains (lower impacts) from export while the net addition of transportation impacts may be ambiguous, but in average conditions the home country will gain in relation to impacts. However the same country can be expected to lose VA, at least in average conditions, although we cannot exclude situations when exporting raw waste rather than treating it domestically will deliver additional VA for the home economy.

The destination country unambiguously loses from the additional environmental impacts of management and loses from the additional transportation impacts inside the country and neither would have occurred without trade. The same country unambiguously gains net additional VA from additional industrial treatment of waste.

In short, while the conclusions for the destination country are unambiguous (environmental loss and VA gain), the conclusions for the shipping or home country are reversed (environmental gain and economic loss) but are more ambiguous and depend on conditions. There is ambiguity also in relation to the net gains and losses from IWT for the two countries together: there may be a net environmental loss coupled with a VA gain or even a VA loss, or a net environmental gain coupled with a VA gain (or loss).

It should be noted that these conclusions arise from a simple *ex post* 'accounting' approach to environmental impacts and VA impacts associated with IWT. A more complete *ex post* analysis of the dynamic economic and environmental impacts of IWT (e.g. additional industrial transformation of the waste flow, impacts on local development and innovation, different transportation modes, waste criminality, etc.) could be expected to produce even more uncertain or ambiguous results.

Note also that rational or profit maximizing choices by collectors to ship the waste abroad for treatment do not necessarily give rise to the optimal social outcome in terms of VA creation and environmental costs to the home country. Suppose, as in the example in footnote seven, that the price of the waste-derived material is the same in the two countries; then there is no VA advantage from trading waste either for the two countries together or for the home country (treating the waste domestically or exporting the waste would give rise to a net VA of 75 along the chain). However, following our example, the home country collector will derive a net VA of 50 from exporting the waste and of 25 from delivering it to a domestic facility. Therefore, it is likely that the waste will be exported by the collector, that is, the owner of the waste. Transportation costs can reduce the profitability of trading if the foreign facility is more distant than the domestic facility (which may not be the case) but the probability of trading will be high, and will be so whenever the collector's VA (and more specifically the profit component of VA) is higher from trading than from domestic treatment, even when the home country economy as a whole loses from trading (i.e. the loss of VA for home country treatment can be often higher than the price advantage from the waste input being delivered to the foreign facility, see Equation (13)). Furthermore, although we cannot be sure that trading will reduce environmental impacts in the home country, the decision of the collector to trade does not incorporate external environmental costs. On the other side, the demand for imported waste inputs (an additional demand in our case) from the foreign treatment facility, which is profitable in terms of industrial and national VA creation, will unambiguously create an external environmental cost to the destination country that will not be taken into account by the importer. In short, the economically optimal individual decision to trade of the home country collector and the foreign treatment facility can be suboptimal for both VA creation and value destruction through environmental impacts. This conclusion puts the focus on the drivers of IWT (see next section).

Finally, we note that the information requirements for full applicability of the approach proposed are huge and not fully satisfied by the available data. For the environmental impacts, in addition to the coefficients from the now numerous LCA datasets and some specific work (Gentil *et al.* 2009), a possibly useful source for transportation impact analysis is the COPERT model developed by the EEA and JRC, among others, to estimate emissions from road traffic.[9] A source of interesting information on certain impacts is provided by E-PRTR. The dataset on emissions from all waste facilities with a reporting obligation is available on the web.[10] However, only data on emissions from individual waste facilities are available. This makes it impossible to estimate the ratio of waste impacts to waste inputs and hence to obtain an indication of whether there are differences in the impact of each unit of waste for each country.

On the economic variables for the waste sector, the European picture is poor, although some countries have good documentation. A new and significant source of data on prices and taxes in the waste sector is the recent work of the Bio Intelligence Service (2012), which presents a collection of information on single EU countries.

A more ambitious objective might be to address other kinds of impacts such as disamenity, transport congestion, etc., which are not extensively documented. However, there is a vast literature on the economic value of environmental impacts of waste management, for example, landfill, and work on external costs of transportation quantified in monetary values, for example, the DG TREN project ExternE (see Mazzanti and Montini 2009 for a set of references). A new set of data on estimated external costs of air pollution and GHGs from industrial sources is presented in EEA (2011).

Drivers of international waste trade

This analysis addresses the question: *Why is waste shipped internationally and not treated domestically?* This question is related to the choices and behaviours of waste producers/collectors, and can be logically separated from the *ex post* environmental and economic impact analyses presented earlier in the chapter (i.e. the consequences of choices already undertaken) unless there are demonstrable links between the drivers of shipments and environmental impacts, for example, the deliberate intention to avoid impacts in the home country.

The policy relevance of this issue is linked to the possibility that, if better understood, the drivers of IWT may be influenced by policies/legislation designed to reduce the negative net impacts of waste trade (if any) or exploit the positive impacts (if any).

Suggestions from the literature

The literature on waste economics and policy includes some methodological references for modelling the drivers of IWT and may provide suggestions about possible analytical approaches.

The general literature on IWT drivers and management/disposal is presently highly biased towards East Asia case studies. At the world level, the United States is the major exporter of waste and China is the major importer of waste. However, EU level analyses have worldwide relevance since Germany, UK, France, the Netherlands and Belgium are among the ten largest world exporters of waste (total waste), and Germany, Spain, Italy, France, the Netherlands and Belgium are among the ten largest world importers of waste. Apart from the UK, all the main countries involved in waste trade are both exporters and importers; the southern countries seem to have higher levels of imports than exports.

Kellenberg (2010) notes that differences in economic, institutional, structural, country-specific regulatory factors may explain waste trade. He stresses the importance of structural (population density), market price (gate fees), and technology/capacity factors, as well as regulatory stringency and enforcement. Cost/price and technology/capacity factors, along the lines of the debate on the 'pollution haven hypotheses', can drive waste trade in the same direction or balance one another.[11] In general, lower management/disposal prices and higher capital intensity (i.e. incineration, recycling sites) – reflecting economies of scale and comparative advantages in recycling and disposal – should attract waste flows. In fact, when dealing with trade issues, 'relative' rather than absolute factors are relevant.[12] At the empirical level, using international COMTRADE data for 92 countries and bilateral trade flows in hazardous and non-hazardous waste, Kellenberg (2010) finds that waste imports increase for a country whose environmental regulations are less stringent vis-à-vis the trading partner, implying a sort of 'waste haven hypothesis'.[13]

International trade research often exploits 'gravity models'.[14] The 'gravity equation' framework constitutes a theoretically and statistically robust basis for analysing the impact of structural bilateral country factors, public policies and innovation on export dynamics (Costantini and Mazzanti 2012).

A study by Baggs (2009) analyses international trade in hazardous waste using a gravity model that includes country characteristics. It concludes that a significant 'pollution haven' effect can be observed: rising per capita income reduces the amount of hazardous waste countries import. This effect is outweighed by high-income countries' relative capital abundance, and by the fact that higher GDP creates larger disposal capacity than waste production. In short, national technology/capacity intensity attracts imports of hazardous waste.

A theoretical and empirical paper by Higashida and Managi (2008) produced evidence denying the 'pollution haven effects' found by Baggs (2009) and Kellenberg (2010). All these works exploit gravity models and different datasets. They use IMF bilateral waste trade data for five kinds of waste and scrap with established international markets, and look at specific evidence using the manufacturing wage as a proxy for development/policy stringency. They show that the more developed the country, the more recyclable waste it imports: the recyclable sector, in their view, is not separated from the final goods industries and consumption destinations. Recycling sectors do not expand faster in less developed countries. In certain circumstances, this evidence does not contradict Baggs (2009). 'Pollution haven' tests within the EU should test whether flows are directed towards the eastern EU and southern EU for reasons related to relatively less stringent enforcement of policy, lags in implementation of waste policies, and generally more lax environmental policy reflected in market prices and non-market variables.

Several scholars have investigated the huge transboundary movements of certain types of waste among Asian countries. They highlight the importance of drivers such as cost and treatment capacity. Fuse and Kashima (2008) develop a very detailed international input-output model to track the automobile recycling systems in Japan and Thailand and the international flows of parts and materials from end-of-life vehicles. Recycling loops of material between Japan and China are analysed in several studies, generally without a specific model structure. Informal trade in waste in Vietnam, linked to Hanoi's rapid economic and spatial change, is analysed in Mitchell (2009). Yoshida and Kojima (2008) analyse transboundary movements of recyclable waste in Japan, focusing on the management system and practices, and case studies reviewing legal frameworks and regulatory implementation. This literature is mostly overly focused on case studies and exchanges within very limited (urban) areas. It consequently offers very limited insights and hints from a conceptual and methodological point of view.

Hints for understanding international flows emerge from analyses at the within-country (region) level. De Jaeger and Eyckmans (2010), studying Flanders municipalities, finds that for some waste – bulky household refuse, demolition waste and garden waste– the quantities collected at the local recycling centre depend on the prices charged at the recycling centres in neighbouring municipalities.

Within-country trade can be influenced by the degree of spatial differences on waste performance across regions in the country. The evidence is highly country-specific and depends mostly on institutional and economic factors influencing the structure of waste regulation from centre to periphery. Ham (2009) finds some degree of convergence for municipalities for recycling, and robust spatial correlations (local areas behave similarly). At a higher administrative level (provinces), Mazzanti *et al.* (2010) find some signs of convergence for waste trends, but also find negligible and decreasing spatial correlations (similarity) over time. Policy decentralization, linked to different economic development, can generate situations where waste management patterns differ, and where waste management specialization in regions and states differs.

A stream of operation analysis models address the optimal distribution of a waste management system (regions/country) that includes the cost structure of managing waste in different facilities, taking account of location and then transportation costs (Ley *et al.* 2002 for the United States). The factors underlying the choice of management site for hazardous waste, including empirical analysis of halogenated solvents in California, is analysed in Alberini and Bartholomew (1999). They conclude that there is a role for public policy which can create different management options thereby influencing the intensity and direction of interstate flows in the United States. However, these models are focused mainly on small geographical areas and are highly demanding in terms of data.

In general, the more that waste trends (production, management) differ and diverge at intra-country and international levels, the more likely it is that waste trade will occur. Heterogeneity drives trade and potentially win-win economic exchanges. Policies should regulate trade and avoid the creation of critical hotspots, particularly those located in densely populated areas, which highlights the issue of environmental impacts.

Discussion and proposals on drivers

Starting from the literature, the factors behind IWT can be discussed in more depth and additional relevant drivers can be proposed. Leaving aside illegal flows, which are not specifically addressed in this chapter, the decision either to ship internationally or to manage domestically is driven by a variety of economic, technological and legal considerations, but we would expect them to lead to (at least expected) *cost saving decisions.* In fact, for a company, the choice between managing domestically or shipping internationally is similar to comparing two projects and their associated costs and benefits.

We propose a set of economic, legal and technological factors that could be relevant to a cost saving decision. Some of these drivers of trade have been acknowledged in the analytical literature and reports (EEA 2009; ETC/SCP 2012). Of course, the importance of single drivers can be very different for different categories of waste. We hereby list the drivers and discuss their role.

1. *Gate fees and transportation costs*. For a given management option, assuming there is enough management/treatment capacity in both countries (home and destination), a basic driver of IWT is the difference between the domestic and foreign gate fees (excluding environmental taxes), taking into account transportation costs. In contrast with the discussion earlier in the chapter, here we refer to gate fees or negative prices (i.e. waste is delivered at a cost) as a usual condition for many waste flows.

   For a given treatment, for example, landfill, of a given waste category, say, MSW, gate fee differences can be significant across European countries as well as across regions within countries, and these differences can stimulate international shipments. However, the differences in gate fees can justify shipment abroad only in conjunction with differences in transportation costs between shipping internationally and managing domestically.

   Formally, to justify export, the advantage (cost saving) in gate fees (assuming negative prices for waste) abroad must be higher (i.e. less negative, or even positive) than the additional transportation costs. In short:

   $$\textit{Export yes, if: } GFH + TH > GFD + TD \textit{ or: } GFH - GFD > TD - TH \qquad (16)$$

   where:

   GFH = gate fee (negative price) at a hypothetical site in home country;

   GFD = gate fee (negative price) in destination country;

   TD = transport cost in destination country;

   TH = transport cost for hypothetical management at home.

   For example, if the gate fee in Germany is 100 and the gate fee in Italy (at the closest facility) is 120, there will be a gate fee saving of 20 (GFH – GFD) for exporting to Germany. However, if the transport to Germany is more expensive than transport to the closest facility in Italy, shipment will only take place if this additional transportation cost (TD – TH) is less than 20.

   Equation (16) is valid even in apparently paradoxically situations created by distance to management/treatment sites. For example, we cannot exclude that gate fees are higher in the destination country (e.g. GFD – GFH = 20), but transportation distances within the country may be so great (e.g. from northern to southern Italy as the closest treatment site at home) and domestic transportation costs so high (e.g. TH – TD >20), that a producer/collector may find it profitable to export. Location of waste produced/collected may be an important factor in creating microeconomic advantages or disadvantages to export.

   In considering transportation costs, we need to take account of the workings of the freight market. The so-called 'return freight', that is, those applied to rent otherwise-empty ships and trucks, may be very low – the alternative being return travel without a load. This may be a factor favouring international shipment for non-hazardous waste and/or all waste categories (e.g. waste paper) that do not require specialized or specific transport facilities and logistics. This applies especially to long-distance shipments, for example, China by ship.

Another factor to be considered is the availability of close international harbours, for example, the Netherlands, which may explain the concentration of waste shipments and transits in some countries. Overall infrastructural assets can attract imports or reduce exports, with exceptions determined by some specific border conditions or large EU infrastructures connecting various countries.

2. *Administrative costs.* For all waste flows, administrative costs must be added to the transportation cost to the destination country, which may reduce the profitability of international shipment. For some waste flows, specific administrative costs for compliance with legal requirements must be added, for example, notification fees. For some waste categories, for example, hazardous waste, these costs may be high (compared to gate fees) and may discourage (legal) international shipment.

   The procedural factors positively or negatively driving waste trade in non-hazardous recyclable materials, especially to non-OECD countries, are set out in OECD (2010) on the basis of case studies (for the EU and the Netherlands) and interviews with waste managers. The net effect of these factors may be uncertain, for example, by implicitly restricting some export/import possibilities with certain countries they may shift trade to other countries. For hazardous waste, factors such as the export ban on non-OECD countries (Articles 34 and 36 of the Waste Shipment Regulation) may be a very important inhibitor of shipment of waste which is necessary to prevent transfer of environmental problems from Europe to these countries. It may shift trade in this same waste towards OECD countries (if other favourable conditions apply).

3. *Tariff and non-tariff barriers.* For non-EU destinations, there may be tariffs on waste in importing countries that may discourage IWT (for a given difference in gate fees and transportation costs). In addition, for certain categories of waste there may be non-tariff customs measures (or even bans) that may increase the costs of international shipment or even prevent it. This may also explain the geography of trade if, outside the EU, destination countries impose different tariff rates or non-tariff customs barriers. Kellenberg (2010) includes 'free trade area' and 'Basel convention ratification' as variables that could introduce geographical-trade policy. The existence of free-trade areas has a plausible positive effect on trade whereas Basel Convention ratification has a negative sign for trade.

4. *Differences in environmental taxes and policy stringency.* There may be differences in environmental taxes charged to the same management/

treatment in different countries, for example, taxes on landfill or incineration. Given the gate fees (and other costs), this may stimulate shipments out from high-tax countries.[15]

Kellenberg (2010) analyses specifically how and whether differences in regulatory stringency drive trade or not. He uses a subjective international level survey-based index. For the EU, both EUROSTAT and EIONET sources provide robust information on country environmental policy stringency (e.g. energy and environmental taxes used in Costantini and Mazzanti 2012). A waste policy index based on EIONET Country Fact Sheets is used in Mazzanti and Zoboli (2009) and Nicolli and Mazzanti (2011) to analyse the role of policy as a driver of waste management, landfill diversion and waste technologies. Other economic instruments and (voluntary) Environmental Management Systems (EMS) can also be considered drivers.

In a transition phase of the regulatory process, different liability rules across EU countries can be used to explain waste trade, in particular waste shipped for disposal. 'Race to the bottom' and 'waste haven' hypotheses may be related to lax liability schemes (Helm 2008). However, the adoption and diffusion of EMS could be positively correlated with higher recovery standards, and EMS diffusion could be a quality asset that attracts waste flows (i.e. Germany). The net effects of EMS may be unclear and its role needs to be evaluated alongside other factors. For example, Iafolla *et al.* (2012) find no correlation between EMS and waste generation and landfill.[16]

5. *Differences in treatment capacity across countries (excess supply/demand).* For a given technology, for example landfill, there may be saturation of capacity at a reasonable distance in the home country and an excess supply of waste requiring management. This may occur particularly in small countries and may stimulate IWT in the search for treatment capacity. In some critical hotspot situations, such as the well-known Campania/Naples (Italy) case, leaving aside illegal market factors, institutional failures and lack of diversification in (technological) options play a significant role (Pasotti 2010a, b). International trade in this case has been and continues to be a necessary, but not always the least-cost solution.

Trade is efficient only if its cost is lower than the best/least cost mix of waste treatments at local level. However, dynamic considerations may be relevant: short-term benefits of trading waste could delay or stop investment in capacity in the exporting region, which will rely on capacity available elsewhere. Thus, it may be that trade is beneficial in the short run, but delays more efficient investment and solutions and induces dependence on foreign capacity. More generally, the transition from traditional disposal in landfill to recycling/recovery as a consequences of the Landfill Directive or early national policies on separate collection of non-hazardous waste (EEA 2009; ETC/SCP 2012), may result in a lack of recycling/recovery capacity in the home country. This may create the conditions for trade if there is enough industrial capacity (in excess of domestic waste supply) in other countries. This may apply also if diversion from landfill reduces the capacity for absorption of the residues from waste recovery/recycling, for example, incineration ash. Lack of domestic capacity and lack of diversification in management/disposal options is one reason for trade that de Jaeger (2010) highlights.

Imbalances between domestic supply and domestic demand (capacity) for different treatment technologies may occur asymmetrically between countries, and change dynamically over time. For example, the development of packaging policy led forerunner countries without the required domestic capacity for recovery/recycling to generate excess supplies of waste/materials. This gave rise to substantial international trade flows, for example, the European/international market for waste paper/paperboard in the 1990s–2000s. The gradual creation of domestic capacity in most countries reduced these flows and/or changed the quality/composition/destination of trade, for example, towards Asia.

Van Beukering and Bouman (2001) analyse the markets for (waste) paper. Using data on Norway and Indonesia, they show that developed countries have high recovery and separated collection rates that result in supplies in excess of domestic demand (recycling demand). This excess supply is exported to developing countries with high rates of imports and utilization of secondary materials. Developing countries often specialize in waste paper processing and becoming net exporters of final products. The situation within the EU may show some differences for richer and poorer countries.

Similarly, the changing balances of domestic supply/demand associated with policies such as those on WEEE (waste of electric and electronic equipment), ELV (end-of-life vehicles), batteries, etc., can create the conditions for waste exports towards emerging large-capacity markets for materials to recycle/recover, for example Asia.

ETC/SCP (2009 and 2012) concludes that transboundary shipments of waste are aimed at looking for better waste management practices, such as mechanical sorting of mixed waste, recycling, composting, anaerobic digestion, and incineration with energy recovery. These waste management processes generate recyclable materials and energy, but they also generate new waste types, which have to be treated. The increase in transboundary shipments of waste therefore is driven partly by the introduction of recycling and recovery requirements in EU Directives. The response to these policy-driven imbalances through international trade may be conditional (again) on different gate fees for recycling/recovery (in foreign destination) compared to landfill (in the home country), taking into account transportation costs. In fact, instead of gate fees, some of these wastes show positive prices that follow the cycle of international prices for virgin materials, for example, industrial metals.

Although economies of scale are relevant for abating management costs through specialization, then attracting domestic and imported waste (i.e. ambiguous effects on trade), diversification is a value for national waste management systems. A less diversified system could generate higher potential export if national alternatives were limited, as in the case of small countries (e.g. Belgium) (de Jaeger 2010).

6. *Different incentives for recycling/recovery (energy).* For waste suitable for different management options, there may be incentives in certain countries that stimulate demand for waste in excess of domestic supply. An example of these incentives are those arising from policies on RES (Renewable Energy Sources), which include waste as feedstock. Incentives for RES (especially electricity-RES) in some EU countries are high compared to other EU and non-EU countries, and this may stimulate international procurement of waste for energy. Based on information on national incentive schemes related to RES, this factor can be integrated into empirical models as an increasingly important driver, especially since the Directive 2009/28 which boosted EU policies for RES (binding 20 per cent of RES in final energy consumption by 2020, differentiated by Member States).
7. *Differences in legislation/classification.* Any differences in formal legislation (or administrative practice) for waste classification and treatment across countries could create incentives to ship internationally, provided that this difference represents a difference in costs. These costs can be compared to other cost items (gate fees, transport, taxes/incentives, administrative and custom costs) to determine whether IWT is profitable or not. However, in certain cases, for example hazardous waste, the difference in legislation may be so huge to justify export on a cost basis. In the EU, homogeneity is high, though some differences exist (i.e. between the western and eastern EU). The OECD (2010) highlights that a lack of harmonization in EU regulation is perceived as a barrier to trade. Complementary reasoning is developed by Karl and Rannè (1999) who discuss centralization and subsidiarity with respect to EU waste regulation.[17]
8. *Need for specific technologies.* For some waste categories, for example certain hazardous waste, there may be limited or nil capacity for treating those specific wastes according to legislation in the home country. This may drive international flows even over long distances and at high transportation and administrative costs.
9. *Geographical characteristics of countries/regions.* Although they may be proxied by some of the drivers discussed above (especially transportation costs), the geographical characteristics of countries/regions may be relevant drivers of IWT. They are also easier to define than transportation costs and may be used as proxies for them in empirical analyses.

Table 6.1 summarizes the drivers identified. We preliminarily attach to each driver an indicator that can be considered representative of the driver in the empirical analysis, with a negative or positive sign. We did not explore data coverage for these indicators; this could be done in specific applied analyses.

The unit of analysis in the table is *the flow between two countries in one direction (export)*, for example, export from Austria to Germany. Single country-to-country flows (rather than total imports or total exports in/to a country) are the more detailed unit of information available. Their explanation may be related to differences between countries for specific drivers, instead of general country characteristics possibly driving total import/export. The empirical indicators for drivers, that is, the difference between the two countries for that indicator, should be suitable for analysing bilateral flows. However, taking bilateral flows as the unit of analysis is unambiguous in terms of the drivers mainly for countries that are only exporters (i.e. for a homogenous good, inter-country differences or comparative advantages should push trade in one direction, even without complete specialization). This same unit of analysis could instead create interpretation problems for (substantial) flows of the same waste category that take place from country A to country B *as well as* from country B to country A, or intra-industry trade (which is the case for many non-hazardous waste). The drivers of this intra-industry trade in the same waste category (e.g. single HS code level) may be mostly microeconomic, that is, differences between specific treatment conditions at facilities, whatever their location, and may be investigated more easily through interviews than by examination of the data only. For example, if we examine the indicators of gate fees and transportation costs we would expect that all flows of a certain waste code go only from Dutch operators to German treatment facilities; however, there may be a German operator who would find it profitable to ship the same kind of waste to a Dutch facility that offers plant-specific contractual advantages compared to the German facility.

Evolving drivers: suggestions from trade in waste wood

International trade in wood waste has increased rapidly. In Europe, wood waste trade for recycling by the particle board industry has been well developed for a long time. Recent developments of RES policies in European countries that use wood biomass and wood waste as feedstock for electricity and heat production are having a major impact. Wood materials are heterogeneous (wood waste, wood residuals from the wood working industry classified as waste, wood chips, particles, etc.)[18] and subject to complex substitution and complementarity in use.

There are significant problems in tracking wood waste and residues, in both production and in international trade. The classification of these wood

Table 6.1 Driver of bilateral flows (the sign of the indicator is the relevant one in driving export)

Driver	*Indicator (− = difference between …; * = multiplied by …)*	*Expected sign on trade (+ or -)*	*Notes/examples*
Gate fees and transportation costs	GF in exp country – GF in imp country (technology of treatment in imp country)	+	The technology of treatment in imp country is taken as reference assuming the treatment would have been the same in exp country
	Distance between exp and imp countries * average transportation cost	–	Standard distance between geographical centres multiplied by the average transportation cost of the prevailing transportation mode for waste
Administrative costs	Cost of exp/imp practices	–	
Tariff and non-tariff barriers	Existence of bans	–/+	The existence of bans may prevent trading with certain countries (-) but may re-direct the flow to other countries (+), or bans may be relevant for the direction of trade
Differences in environmental taxes and policy stringency	Landfill tax in exp country – landfill tax in imp country Or: Indicator of env. policy stringency in exp country – indicator in imp country	+	Indicators of env. policy stringency available from World Economic Forum

Table 6.1 (cont.)

Driver	*Indicator (– = difference between …; * = multiplied by …)*	*Expected sign on trade (+ or -)*	*Notes/examples*
Difference in treatment capacity (policy-driven excess supply/demand)	Capacity in exp – capacity in importing (treatment in importing)	–	The technology of treatment in imp country is taken as reference assuming the treatment would have been the same in exp country
	Or: (Collection – capacity RR exporting country) – (Collection – capacity RR exporting country)	+	Focus on RR. The higher the excess supply domestic compared to country of destination, the higher the probability of observing bilateral trade flow
Different incentives for recycling/recovery (energy)	Incentive on e-RES in exp country – incentive on e-RES in imp country	–	
Differences in legislation/ classification	Stringency legislation in exp country – stringency legislation in imp country	+	To be elaborated by taking legislation in a country as benchmark
Need for specific technologies	Availability of X technology in importing country only	+ if available	Relevant mostly for hazardous waste
Geographical characteristics of countries region	E.g. common borders	e.g. + if the two countries have a border in common	Sign different for each characteristic

materials in legislation and in administrative practice is not clear-cut and also varies from country to country; it is also not well established in statistics produced by international organizations. For example, in Italy, residues of 'virgin' wood from the furniture industry are classified as 'waste' because of alleged contamination; particle board plants have the status of waste treatment plants (recyclers) allowing them to receive wood from the network of post-consumer collection platforms (linked to Consorzio Nazionale Imballaggi (CONAI), the Italian compliance scheme for packaging policies). Traditionally, these materials have unclear markets, with uncertainties in product definition and legal definition (waste/non-waste), and limited availability or clarity of data.[19]

In international trade statistics, different data sources that use different classifications exist for wood waste. The United Nations Food and Agriculture Organization (UN FAO) classification differs from the classification of the UN's COMTRADE, which is the basis for of EUROSTAT's COMEXT. In the FAO classification of forest products there is no category for 'waste wood' and the relevant products in the FAOSTAT and FAO trade flow database are 'Chips and Particles' and 'Wood Residues', which *might* include waste.[20] Unfortunately, the FAO classification is not immediately linked (or explicitly trans-codified) to the COMTRADE and EUROSTAT classifications. The COMTRADE classification includes wood waste and 'wood in chips and particles' within the three-digit code 246 ('Wood in chips or particles and wood waste') in the Standard International Trade Classification (SITC) Revision four. COMTRADE also includes 'wood waste' in the four-digit level code 2462 together with sawdust and wood scraps.[21] This category of materials (Wood waste, sawdust and wood scraps) is code 440130 in the Harmonized System (HS), which is also used by EUROSTAT for international trade statistics.[22]

Trade classifications seem to confirm the significant substitutability between wood waste and other categories of wood residues from forestry and industrial operations, both for energy and industrial (recycling) use.

General confirmation that wood waste and wood residues are increasingly crossing the borders of European countries is indicated by the aggregate European import and export of 'wood residues' (FAOSTAT classification and data). There has been a huge increase in trade since the 1990s, from around three million/m^3 to about 14 million/m^3 (export) in 2009 (Figure 6.2). Since 2003, imports have increased more than exports: net imports in 2009 were around five million/m^3. This trend suggests increasing dependence of Europe on wood residues and wood waste from non-European countries.

In European trade in wood waste, a pivotal role is played by Germany as an importer and exporter (between Belgium, the Netherlands and Austria), and of Italy as a major importer of these wood materials for industrial recycling. Trends towards increasing trade (code HS 440130) have been emerging in the cases of Germany and Italy since the 1990s. Up to the early 2000s, Germany was a net exporter of waste wood with an overall flat trend for

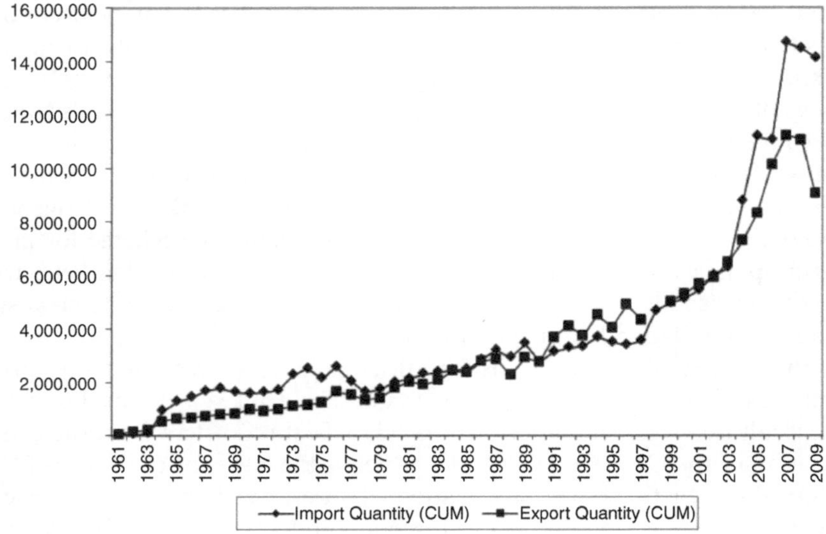

Figure 6.2 Total import and export of 'wood waste/wood residues' in the EU27 (including intra-EU trade, m³).

Source: Own elaboration on FAOSTAT data.

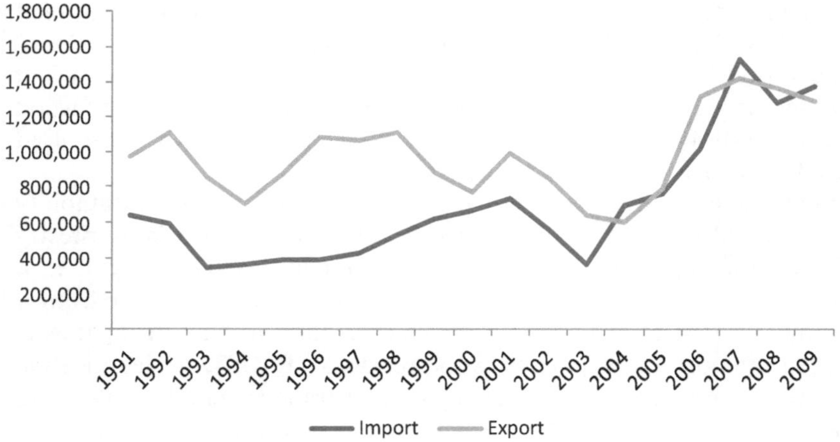

Figure 6.3 Germany, trade of waste wood (HS 440130, tons).

Source: Own elaborations on COMTRADE data.

both export and import (Figure 6.3). After 2003, both imports and exports increased until the present crisis, with imports increasing more than exports. Imports passed from 400,000 tons in 2003 to about 1.4 million tons in 2007. In 2009, Germany is a net importer of wood waste.

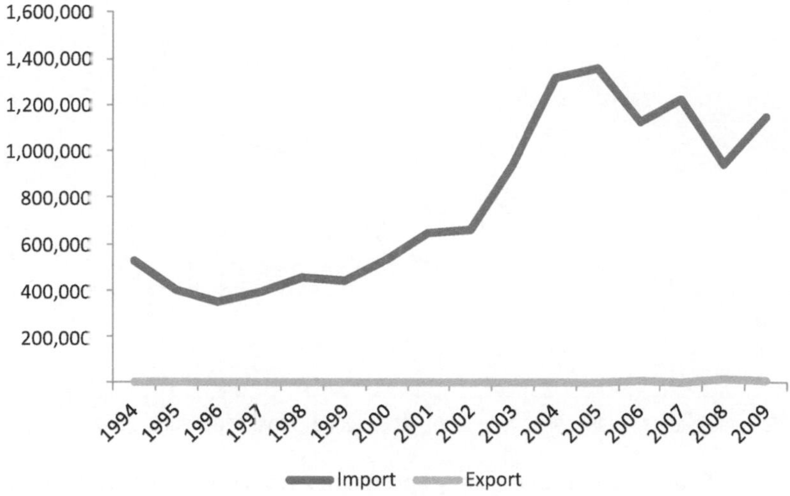

Figure 6.4 Italy, trade of wood waste (HS 440130, tons).
Source: Own elaborations on COMTRADE data.

Italy is structurally a net importer of wood waste (mostly for the particle board industry), and its total imports steadily increased from 400,000 tons in the early 1990s to 1.4 million tons in the mid-2000s (Figure 6.4). The decrease after 2005 may be the result of trends in industries using wood panels.

The observed increase in domestic demand and international trade in wood waste/residues is generally considered to be the result of European policies on RES. International trade flows to and from European countries were well developed for particle board production before the EU Directive 2001/77/EC on the promotion of electricity based on RES (Alakangas *et al.* 2002), and the Directive 2009/28 on the promotion of use of energy from RES. National RES policies and the expectations associated with the 20 per cent target of EU policy on RES greatly increased consumption of wood materials, especially for energy, and complicated the network of commercial flows. Figure 6.5 shows the increase in energy produced from solid biomass (mostly wood and wood waste) in the EU27 from 1995 to 2008. Major increases took place in the production of electricity from solid biomass, especially after Directive 2001/77/EC.

The need to recover all sources of wood biomass contributed in some countries, such as Germany, to a de facto ban on landfilling of wood.[23] Germany has recorded a continual increase in electricity production from solid biomass, which may explain in part the strong increase in imports of wood particles and wood waste. This may be compatible with the observed increase in exports of wood waste and wood residues, given that different kinds of wood waste and residues can have different values in different

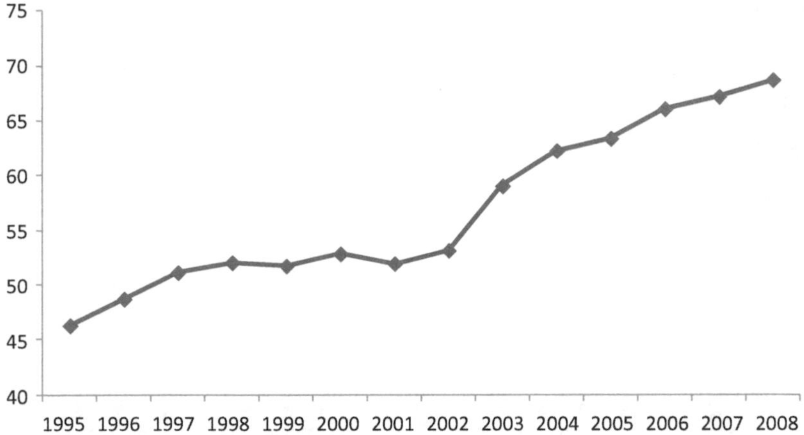

Figure 6.5 Evolution of primary energy production from solid biomass for the EU27 since 1995 (in million tons of oil equivalent).

Source: EurObserver, Solid Biomass Barometer (2009).

industries. There has been a decrease in wood exports from traditional export/transit countries, particularly the Netherlands, because of the rapid development of domestic demand for wood for energy production.

In Italy, the production of electricity from solid biomass (mostly wood) increased fivefold between 2000 and 2008. There has been a major increase in the consumption of pellets (produced from wood dust), which has induced an important inter-Alpine international market. However, industrial demand (recycling) of wood residues and waste also increased in Europe up to 2006/2007, before the crisis. In many countries, there has been a shortage of wood residues and other biomass suitable for energy production and industrial recycling, with strong competition among different uses and increasing prices for these materials.

Together with the increasing demand of wood for energy, there has been a continuous penetration of particle board from wood recycling in the wood material markets. Figure 6.6 shows the ratio between apparent consumption (production + import – export) of particle board and apparent consumption of sawn wood for the EU27, indicating the relative market shares of virgin and recycled wood. The ratio has increased steadily from less than 10 per cent in the 1960s to 70 per cent in 2009.

The demand for wood for energy, which has partly displaced the wood panel industry, is raising questions about whether RES policies may be causing an 'inversion' in the EU 'waste hierarchy' that is favouring energy recovery over recycling of waste materials. The woodworking industries have highlighted this problem and are proposing a more balanced approach that takes account of the GHG impact of using waste wood in industrial

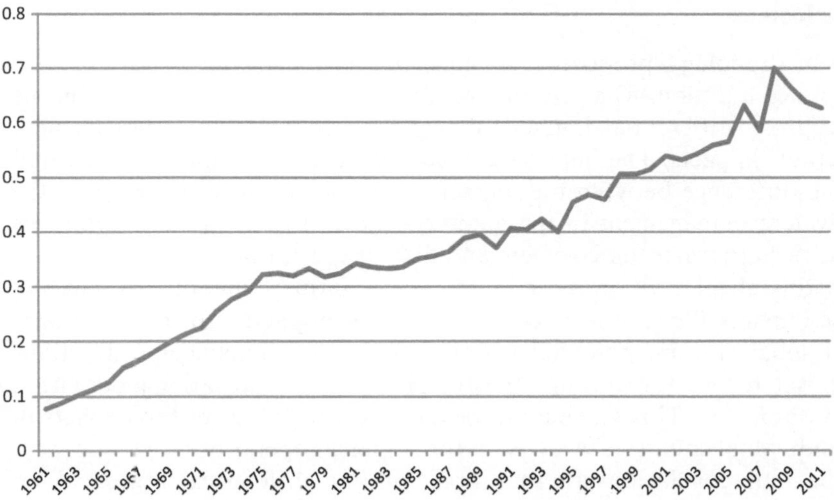

Figure 6.6 Ratio between the apparent consumption of particle boards and sawn wood in the EU27, 1961–2011 (percent).

Source: Own elaboration on FAOSTAT data.

mechanical recycling plants as well as its use in energy biomass plants. Unlike energy recovery uses of wood, mechanical recycling does not provide immediate release in the atmosphere of the carbon content of wood (CEI-Bois 2006; Eggers 2002).

The two main sources of demand (industry and energy) were both booming before the present crisis, and operators were searching for additional sources of supply in other countries, giving rise to trade flows in multiple directions (UNECE 2006). The exhaustion of easy-to-access sources of wood materials is making Europe a net importer of waste wood and wood residues from the rest of world.

In short, the drivers of additional wood trade have been:

- different incentives for RES policies in different countries, and then favourable prices (or gate fees) at facilities in certain countries;
- uneven capacity for biomass energy in countries, which may change with changes in national RES policies towards the targets in the RES Directive 2009/28;
- structural differences in the development of the particle board industry across countries (recycling), and differences in the short-term business cycle of this industry in different countries;
- geographical proximity to unexploited sources in neighbouring countries (transportation costs);
- environmental motivations playing a limited role in trade.

Conclusion

The methodology proposed for analysing *environmental impact ex post* (reference question: what are the impacts of waste trade?) is based on the concept of 'differential impact'. Wherever it occurs, waste management generates impacts. The impacts attributable to IWT should be considered as the difference between the impacts from management abroad and the equivalent management in the home country. In both cases, the impacts come from (i) waste management and (ii) transportation.

In this chapter we proposed simple accounting equations to measure these impacts. This impact accounting can be applied to different categories (global warming potential, toxicity, etc.), different management options (disposal, recovery, recycling, etc.) and different transportation modes (rail, road, ship, etc.). This scheme can be adapted to analysis of the impacts in the individual countries involved in the shipment operations.

The main conclusions are that, first, within a 'differential approach', we cannot state *ex ante* that the export operation is always a source of overall *net* environmental impacts compared to national treatment; in the case of short distance, cross-border shipments and more efficient technologies in the destination country, we cannot exclude *ex ante* that IWT will have an overall net beneficial impact; however this possibility may increase if the IWT is associated with a shift up in the 'waste hierarchy', for example, from landfill disposal at home to recycling abroad; net overall environmental impact then becomes an empirical issue that should be considered case by case. Second, when considering the separate expected environmental impacts in the two (or more) countries involved, the shipping country unambiguously gains management impacts while the net addition of the impacts from transportation may be ambiguous: high if management is available at a short distance within the country, or even negative if the domestic management site is very far from the site of production/collection; in average conditions, we would expect the home country to gain in terms of environmental impacts; the destination country instead loses from the additional environmental impacts of management and loses from the additional transportation impacts within the country, because neither would have occurred without the IWT; then the destination country unambiguously loses (although it may gain in relation to industrial activity).

In this chapter we proposed a scheme for analysing the net gain/loss of value added from trading a given flow of waste rather than treating it in a domestic facility (differential approach). The main conclusions are:

- net creation of VA from trade for the two countries together will take place only if the price of the waste-derived material after treatment in the destination country is higher than the price of the same material in the country where the waste originated; these results depend on the features of the VA as a 'consolidated account' across vertical chains of production;

- the home country will unambiguously lose treatment VA in the case of export but may enjoy additional VA from exporting the waste if the price for waste as an input, in the country of destination, is higher than the price of the same waste in the domestic waste industry; although we cannot exclude this possibility, given the huge differences among materials industries across EU countries, it is likely that the home country economy will lose VA from trading its waste;
- the destination country will *unambiguously gain VA from import*, the gain being the difference between the domestic price for waste-derived material and the price of the waste input coming from abroad, under the assumption that this waste input is additional and not a substitute for a unit sourced domestically in the destination country; the gain of VA of the destination country will go entirely to the treatment stage.

If the conclusions on environmental and VA impacts are merged, while the outcomes for the destination country are unambiguous (environmental loss and VA gain), the conclusions for the shipping or home country are reversed (environmental gain and economic loss), but are more ambiguous and depend largely on specific conditions. The same ambiguity applies to the net gains and losses from IWT for the two countries together: there may be a net environmental loss coupled with a VA gain or even a VA loss, or there may be a net environmental gain coupled with a VA gain (or loss).

The proposed approach to analysing the drivers of IWT (reference question: what drives the decisions to trade waste?) is based on the discussion of a set of economic and institutional drivers only partly considered in the literature. The decision to ship waste internationally or to manage it domestically is actually driven by a variety of economic, technological and legal considerations, but it is expected to lead to *cost savings* (or increased profits). In fact, the choice between managing domestically and shipping internationally is similar to comparing the costs and benefits associated with two projects.

The major drivers of cost-reducing (profit-increasing) IWT are:

- differences in gate fees (e.g. less negative gate fees or positive prices in the export country than in the home country);
- transportation costs (e.g. international transport is less expensive than long-distance transport at home);
- administrative costs (e.g. cost of export/import practices; existence of bans in non-OECD countries);
- differences in environmental taxes and policy stringency (e.g. an incineration tax in the exporting but not the importing country);
- tariffs and non-tariff barriers at borders for non-EU shipments;
- differences in treatment capacity (e.g. less capacity in the exporting country than in the importing country);
- different incentives for recycling/recovery (e.g. lower incentives to produce energy from waste in the exporting country than in the importing country);

- opportunity for more profitable treatment in the destination country;
- differences in legislation/classification (e.g. stricter legislation in the exporting country than in the importing country);
- need for specific technologies (e.g. availability of a certain technology only in the importing country);
- geographical characteristics of countries and regions (e.g. island, small country, long borders, short distance to a facility in the destination country, etc.).

These drivers seem to be relevant to different extents for both hazardous and non-hazardous waste shipped internationally. In the case of non-notified waste, for example, waste of electric and electronic waste, other drivers may be relevant, for example, high dismantling costs in the home country. Some of these drivers of trade are acknowledged by the EEA (2009).

The case of wood waste trade highlights the role of imbalances in the capacity of the particle board industry (wood recycling) across countries and the role of RES to boost international demand and circulation of wood waste. More in general, developments in industrial recycling and energy recovery, which take place at different at speeds and degrees of specialization in different countries, are at the core of trade flows, with environmental motivation playing only a minor role, if any.

Acknowledgements

This work is based partly on analyses developed by the authors at ETC/SCP, European Topic Center on Sustainable Consumption and Production, supported by the European Environment Agency. The authors are grateful to Susanna Paleari and Massimiliano Volpi for providing data and information for a previous version of this work. The usual disclaimer applies. This work also includes some preliminary results from the Work Package 8 (Waste and recycling) of the 7FP project EMInInn – Environmental Macro-Indicators of Innovation.

Notes

1 Notified waste includes hazardous waste, mixed household waste, residues from the incineration of household wastes and other waste categories that have to be notified before shipment according to the Waste Shipment Regulation (EC 2000, 2006). Hazardous waste represents a large share of notified waste to a total of 3.2 million/tons in 2001 and 7.2 million/tons in 2009. The trade in notified waste does not represent all trading flows of goods classified as waste according to waste definitions and lists, and the EU waste trade as a whole is much greater than EU notified flows. Analyses of transboundary shipments of waste, largely based on notified flows, have been developed by ETC/RWM (2008) and ETC/SCP (2009). A more extensive analysis, to which the authors have contributed, is presented in ETC/SCP (2012).

2 Assuming that the back travel of an international shipment of waste is not empty, the impact of the shipment itself does not include back travel.

3 It can be adapted also to the case where impacts are non-linear with respect to the quantity of waste.

4 Long distance transportation of waste may increase environmental risks compared to short distance transportation. For a given category of waste, there can be two main sources of risk: (a) accidents in transport, with related spills into the environment; (b) risk of illegal manipulation on the waste flow, in particular the possibility that waste are illegally passed from one category to another, e.g. by falsifying documentation and/or by intermediate operations of waste mixing. Both risks have a probabilistic implication in terms of higher environmental impacts. Different categories of waste may present different degrees of both risks. To simplify, we can assume that, for each category of waste, both kinds of risks can be associated with the distance travelled, and then, in a perspective of net impact of IWT only, to the net additional transportation distance compared to home management (which may be even negative), or: R = f(TD – TH), where R = risk of impacts in transport; TD = Distance travelled for international shipment, TH = Distance travelled to a hypothetical home destination. The function f(.) may be non-linear, i.e. increasing more than proportionally for increasing distances.

5 If we include risk in proportion to the distance travelled (and then, the net additional travelled distance compared to home management), the distribution is asymmetric between the two countries. The increase in risk in the home country is ambiguous as is the net additional distance within the country. Instead, the increase in risk is a net increase in the destination country, in proportion to the distance travelled in the country – which would be zero for no international shipment. Assuming the same function of risk in both countries: RH + RD = f(TB –TH) + f(TBD) = f(TD – TH) = R, where RH = risk in home country; RD = risk in destination country; TB = distance from the border, home country; TH = distance traveled for management inside the border, home country; TDB = distance travelled from border to management site, destination country; TD = TB + TBD; R = total risk.

6 We assume VA per km is the same for transportation domestically and abroad (VATkm). Defining VATkm as the difference between the unit (per km) price of service (TCkm) and the unit (per km) intermediate input cost (ITCkm), then total VA for domestic transportation can be: TCH –ITCH = DH * (TCkm – ITCkm) while total VA for transportation abroad will be: TCD – ITCD = TCD – ITCH = DD * (TCkm – ITCkm). The difference between the two depends only on the difference between the two distances, i.e. to the home country treatment facility DH and to the foreign facility DD.

7 An example can be used to clarify this point. Suppose the price of the material produced by treating 1 ton of waste, e.g. secondary aluminum, is 100 in the two countries (PD = PH = 100) with the same productivity of transformation. Suppose the price of 1 ton of the waste (aluminum scrap) used as input by the treatment facility is 50 in the home country, e.g. Italy (PWH) and 75 in the possible destination country for treatment, e.g. Germany (PWD). Assuming that waste is the only input, the VA in treatment will be 25 in Germany (PD – PWD = 100 – 75) and 50 in Italy (PH – PWH = 100 – 50). The collector in the home country has a collection cost (ICH) of 25 whatever the destination of the waste: therefore it will have a VA (here, gross of transport costs) of 25 for delivery in Italy (PWH – ICH = 50 – 25) and 50 in Germany (PWD – ICH = 75 – 25). Therefore by choosing domestic treatment, the total VA in the chain will be 75 (25 at the collector stage and 50 at the treatment stage); by choosing to export, the VA in the chain for the same waste will be 75 (50 at the collector stage and 25 at the treatment stage), i.e. the same as domestic treatment. Therefore, if we take the *difference* in total VA for domestic treatment and for treatment abroad it is exactly zero as is, in our example, the difference between PD and PH, i.e. 100 in both countries. If we introduce .

transportation VA and cost, which by assumption is allocated to the home country, the result does not change. Suppose the cost of transportation to the domestic facility is 10 (TCH); this must be added to the cost to the collector thus decreasing its VA by 10 but creating VA in the transportation sector. Assuming a cost for intermediate input in transport (ICTH) of five, the VA created in transport will be five (TCH – ICTH). Since the VA of the collector decreases by 10 and the VA for transport will be five, the net VA for whole chain of treatment at home will be: 75 – 5 = 70. In the case of export for treatment in Germany we can assume a longer distance travelled by the waste (but this will not necessarily be the case) and a transportation cost abroad (TCD) of 15. Assuming again that the transport cost is allocated to the home country collector, this actor will see its VA decreasing by 15. However, if the intermediate costs of transportation are the same (ICH = 5), the VA for the transportation sector will be 10 (TCD – ITCH). Therefore the net VA in the chain for export will be 75 – 5 = 70, which is the same as in the domestic treatment chain. Again, the total VA *difference* between treating domestically and exporting for treatment will be zero for the two countries *together*, or the same as the difference between the price of the waste-derived material at home and abroad (100 in both cases and then PD – PH = 0). Of course, if the prices of the waste-derived material are different, e.g. PD = 120 and PH = 100, the net VA from trading instead of treating domestically will be 20, equivalent to a VA accruing to the treatment facility abroad but also to an overall 'gain from trade'. The example shows that while the specific VA for different stages can differ and VA distribution between the two countries can differ, which can be relevant to the decision to trade or not (see next sections), the net *ex post* VA gain from trade depends, under our assumptions, only on the difference in the final prices of the waste-derived material in the two countries.

8 At first sight it would seem that the country should enjoy transportation VA (assuming that the transportation company belongs to the home country) but the additional transportation VA for exporting instead of delivering to a domestic facility (assuming the distance to the facility abroad is greater) will be an additional cost, and then a reduction in the VA for the domestic collector.

9 See EEA (2000), www.eea.europa.eu/publications/Technical_report_No_50.

10 See www.eea.europa.eu/data-and-maps/data/member-states-reporting-art-7-under-the-european-pollutant-release-and-transfer-register-e-prtr-regulation.

11 The 'pollution haven hypothesis' suggests the possibility that 'dirty' industries, in this case waste management, can move to countries where environmental regulation is less stringent.

12 In empirical trade analysis, this is modelled by a 'gradients' variable that captures the relative strength/intensity/value of bilateral country factors.

13 He notes that a relevant part of hazardous waste may be not reported to the Basel Convention, and instead can be found in the COMTRADE harmonized system as non-hazardous waste. Also referring to Baggs' (2009) figures, he claims that the share of non-reported hazardous waste can be substantial.

14 Gravity models explain the structure of trade according to a few macro variables that can create trade attraction between countries. The simplest models include GDP of countries involved in trade and their geographical distance.

15 A potentially interesting data set on environmental taxes is provided by EEA-OECD (www2.oecd.org/ecoinst/queries/index.htm). It provides information on the qualitative aspects of waste taxation, similar to that available on EIONET. An interesting new source is Bio Intelligence Service (2012).

16 Data on EMS in EU countries are available from EUROSTAT.

17 'If no consensus can be reached about harmonization, should member states be allowed to stop cross-border shipments of waste, or should the EU strive for a common market for waste? We take the position that most objections against

waste shipments are not convincing, especially if the member states implement minimum standards for landfills and waste processing facilities and common information and control systems. Competition between different national regulations within adequate constraints that, e.g., control transboundary externalities leads to more efficient waste management structures in Europe than national self-sufficiency or centralized decision-making at the EU level' (Karl and Rannè 1999).

18 Here we address 'notified' and 'non-notified' flows of wood waste. According to industry sources (interview with the environment and safety specialist of Federlegno-Arredo, the Italian association of wood working industries), the distinction between 'notified' and 'non-notified' waste is based solely on the origin of the waste rather than on substantial differences between materials and/or their industry destination. Wood waste from construction and demolition and waste management facilities is subject to notification because of its origins, which renders it 'contaminated' or, at least, outside the green list. Similar wood wastes from other operations are included in the green list and do not need notification. Wood industry operators suggest that both notified and non-notified flows may be similar materials that could be directed to the same recovery or recycling operations. Therefore, from a statistical and industrial point of view, notified flows may constitute part of major flows of 'waste wood', including non-notified waste and, partly, virgin wood residuals.

19 For the classification of wood materials in the framework of RES policy in Italy, see Paleari (2007).

20 '1619 Chips and particles: Wood chips and particles: Wood that has been deliberately reduced to small pieces from wood in the rough or from industrial residues, suitable for pulping, for particle board and fiberboard production, for fuelwood or for other purposes. 1620 Wood residues. Miscellaneous wood residues: Wood residues which have not been reduced to small pieces. They consist principally of industrial residues, e.g. sawmill rejects, slabs, edgings and trimmings, veneer log cores, veneer rejects, sawdust, bark (excluding briquettes), residues from carpentry and joinery production, etc'.

21 The description is as follows: sawdust and wood waste and scrap, whether or not agglomerated in logs, briquettes, pellets or similar forms.

22 Code 440130 is the one considered by Kellenberg (2010).

23 German legislation bans landfilling of biological waste with calorific potential, and then of wood waste.

References

Alakangas, E., Hillring, B. and Nikolaisen, L.S. (2002) 'Trade of Solid Biofuels and Fuel Prices in Europe', *mimeo*.

Alberini, A. and Bartholomew, J. (1999) 'The Determinants of Hazardous Waste Disposal Choice: An Empirical Analysis of Halogenated Solvent Waste Shipments', *Contemporary Economic Policy*, 3(17): 309–20.

Baggs, J. (2009) 'International Trade in Hazardous Waste', *Review of International Economics*, 1(17): 1–16.

Bio Intelligence Service (2012) 'Use of Economic Instruments and Waste Management Performances', Final Report, 10 April, Contract ENV.G.4/FRA/2008/0112.

CEI-Bois (2006) *Tackle Climate Change: Use Wood*, Brussels: CEI-Bois.

Costantini, V. and Mazzanti, M. (2012) 'On the Green and Innovative Side of Trade Competitiveness? The Impact of Environmental Policies and Innovation on EU Exports', *Research Policy*, 41: 132–53.

De Jaeger, S. and Eyckmans, J. (2010) 'Do Households Export their Recyclable Waste?', paper prepared for the Association of Environmental and Resource Economists (AERE) World Conference, Montreal 28 June–2 July, special session on waste management in decentralized settings: spatial issues and policy.

Eggers, T. (2002) 'The Impacts of Manufacturing and Utilisation of Wood Products on the European Carbon Budget', Internal Report 9, Joensuu, Finland: European Forest Institute.

European Commission (EC) (2000) *European List of Waste*. Commission Decision of 3 May 2000 replacing Decision 94/3/EC establishing a list of wastes pursuant to Article 1(a) of Council Directive 75/442/EEC on waste and Council Decision 94/904/EC establishing a list of hazardous waste pursuant to Article 1(4) of Council Directive 91/689/EEC on hazardous waste (OJ L 226, 6.9.2000, p. 3).

European Commission (EC) (2006) *Regulation (EC) No 1013/2006 of the European Parliament and of the Council of 14 June 2006 on shipments of waste* (OJ L 190, 12.7.2006, pp. 1–98), http://eur-lex.europa.eu/LexUriServ/LexUriServ.do?uri=OJ:L:2006:190:0001:0001:EN:PDF.

European Environmental Agency (EEA) (2000) *COPERT III Computer Programme to Calculate Emissions from Road Rransport – User Manual*, EEA, www.eea.europa.eu/publications/Technical_report_No_50.

European Environmental Agency (EEA) (2009) *Waste without Borders in the EU*, EEA Report No. 1/2009, European Environment Agency, www.eea.europa.eu/publications/waste-without-borders-in-the-eu-transboundary-shipments-of-waste.

European Environmental Agency (EEA) (2011) *Revealing the Costs of Air Pollution from Industrial Facilities in Europe*, EEA Technical Report No. 15/2011, Copenhagen: EEA.

European Topic Centre on Resource and Waste Management (ETC/RWM) (2008) *Transboundary Shipments of Waste in the EU. Developments 1995–2005 and Possible Drivers,* European Topic Centre on Resource and Waste Management, Technical Report 2008/1, available at http://eea.eionet.europa.eu/Public/irc/eionet-circle/etc_waste/library?l=/working_papers/shipments290208pdf/_EN_1.0_&a=d.

European Topic Centre on Sustainable Consumption and Production (ETC/SCP) (2009) 'Data Availability on Transboundary Shipments of Waste based on the European Waste List', European Topic Centre on Sustainable Consumption and Production, Working paper 3/2009, http://scp.eionet.europa.eu/publications/Transbound%20data%20report.

European Topic Centre on Sustainable Consumption and Production (ETC/SCP) (2012) 'Transboundary Shipments of Waste in the European Union. Reflections on Data, Environmental Impacts and Drivers' (by Fischer C., Junker H., Mazzanti M., Paleari S., Wuttke J., Zoboli R.), ETC/SCP Working Paper 2/2012, European Topic Centre on Sustainable Consumption and Production, Copenhagen, available at http://scp.eionet.europa.eu/publications/wp2012_2.

Fuse, M. and Kashima, S. (2008) 'Evaluation Method of Automobile Recycling Systems for Asia Considering International Material Cycles: Application to Japan and Thailand', *Journal of Material Cycles and Waste Management*, 10: 153–64.

Gentil, E., Clavreul, J. and Christensen, T.H. (2009) 'Global Warming Factor of Municipal Solid Waste Management in Europe', *Waste Management & Research*, 27(9): 850–60.

Ham, Y.J. (2009) 'Convergence or Recycling Rates in the UK: A Spatial Econometrics Perspective', paper presented at the European Association of Environmental and Resource Economists (EAERE) conference, Amsterdam, 24–27 June.

Helm, C. (2008) 'How Liable should an Exporter be? The Case of Trade in Hazardous Goods', *International Review in Law and Economics*, 28: 263–71.

Higashida, K. and Managi, S. (2008) 'The Determinants of Trade in Recyclable Wastes, the Structure of Recycling Sector and the Effects of Trade Restrictions', *mimeo*.

Iafolla, V. Mazzanti, M. and Nicolli, F. (2012) 'Waste Dynamics, Country Heterogeneity and EU Environmental Policy Effectiveness', *Journal of Environmental Policy and Planning*, 14: 371–93.

Karl, H. and Rannè, O. (1999) 'Waste Management in the European Union: National Self Sufficiency and Harmonization at the Expense of Economic Efficiency?', *Environmental Management*, 23: 145–54.

Kellenberg, D. (2010) 'Tracing Waste', paper prepared for the Association of Environmental and Resource Economists (AERE) World Conference, Montreal, 28 June–2 July, special session on waste management in decentralized settings: spatial issues and policy.

Ley, E., Macualey, M.K. and Salant, S.W. (2002) 'Spatially and Intertemporally Efficient Waste Management: The Cost of Interstate Trade Restrictions', *Journal of Environmental Economics and Management*, 43: 188–218.

Mazzanti, M. and Montini, A. (2009) *Waste & Environmental Policy*, London: Routledge.

Mazzanti, M. and Zoboli, R. (2009) 'Waste Generation, Incineration and Landfill Diversion De-coupling Trends, Socio-economic Drivers and Policy Effectiveness in the EU', *Environmental & Resource Economics*, 44(2): 203–30.

Mazzanti, M., Montini, A. and Nicolli, F. (2010) 'Waste Generation and Landfill Diversion Dynamics: Decentralized Management and Spatial Effects', Nota di lavoro 27, FEEM, Milan.

Mitchell, C. (2009) 'Trading Trash in the Transition: Economic Restructuring, Urban Spatial Transformation, and the Boom and Bust of Hanoi's Informal Waste Trade', *Environment and Planning A*, 41: 2633–50.

Nicolli, F. and Mazzanti, M. (2011) 'Diverting Waste: The Role of Innovation', in *Invention and Transfer of Environmental Technologies, OECD Working Paper*, Paris: Organisation for Economic Cooperation and Development, Environment Directorate.

Organisation for Economic Cooperation and Development (OECD) (2010) 'Reducing Barriers to International Trade in Non-hazardous Recyclable Materials: Exploring the Environmental and Economic Benefit', Joint Working Party on Trade and Environment, Paris: OECD.

Paleari, S. (2007) 'La definizione legislative di biomassa rinnovabile', in T. Gargiulo and R. Zoboli (eds), *Una nuova economia del legno tra industria, energia e cambiamento climatico*, Milan: Franco Angeli, pp. 252–86.

Pasotti, E. (2010a) *Political Branding in Cities: The Decline of Machine Politics in Bogota, Naples, and Chicago*, Cambridge: Cambridge University Press.

Pasotti, E. (2010b) 'Sorting through the Trash: The Waste Management Crisis in Southern Italy', *South European Society & Politics*, 14: 199–227.

United Nations Economic Commission for Europe (UNECE) (2006) 'Mobilizing Wood Resources. Can Europe's Forests Satisfy the Increasing Demand for Raw

Material and Energy under Sustainable Forest Management?', Background Paper for the Workshop (Geneva, Switzerland, January 2007), UNECE/Food and Agriculture Organization of the United Nations, www.unece.org/fileadmin/DAM/timber/docs/dp/dp-48.pdf.

Van Beukering, P. and Bouman, M. (2001) 'Empirical Evidence on Recycling and Trade of Paper and Lead in Developed and Developing Countries', *World Development*, 10: 1717–37.

Yoshida, A. and Kojima, M. (2008) 'Transboundary Movement of Recyclable Resources: Current Management Systems and Practices in Japan', in M. Kojima (ed.), *Promoting 3Rs in Developing Countries, Lessons from the Japanese Experience*, IDE-JETRO, pp. 146–71.

7 Do weak environmental regulations determine the location of US exports of SLAB and lead waste?

Derek Kellenberg

Introduction

The integration of international markets over the course of the past two decades has received a great deal of attention in the popular press and the academic literature. While international markets have been connected through trade and investment for centuries, advances in transportation and communication technologies have rapidly advanced the pace at which global markets have become integrated. Offshoring, outsourcing, financial crisis and the effects of trade on development, growth and wages have become primary economic and political issues for many countries. Likewise, differences in environmental standards across countries and their influence on foreign direct investment (FDI) and trade has become an increasingly important topic. In particular, the idea that dirty polluting industries are locating in countries with the least restrictive environmental standards, known as the pollution haven effect, has become a focal concern. Most of the pollution haven literature has focused on and defined pollution havens based on pollution as a by-product of goods production in the country where products are produced. Indeed, many papers[1] that have estimated pollution haven effects have found that countries or regions with higher environmental standards can have a negative and statistically significant impact on FDI and net exports. Although pollution haven effects have been shown to be statistically significant, the overall effects on trade or FDI tend to be relatively small compared to other important factors such as wage differences, income or proximity to markets.

More recently, researchers have begun to look at a different, and potentially more environmentally challenging, form of pollution haven effect. Baggs (2009) and Kellenberg (2012) look at how differences in country characteristics can affect the international trade of hazardous and non-hazardous waste across countries. They find that lower incomes and lower environmental standards have a positive effect on a country's imports of waste, creating a waste haven effect in countries with the lowest levels of income and environmental regulation. Given the rapidly expanding volume of international waste trade in the past decade and the potential environmental consequences of these waste flows, this is an area of research that is increasing in policy importance yet remains largely unexplored in the literature.

This chapter contributes to this literature by examining the effects of differences in the stringency of environmental regulations on a particularly hazardous, but increasingly traded, form of hazardous waste: Spent Lead Acid Batteries (SLAB) and lead waste. Lead acid batteries are most commonly found in automobiles but are also used in emergency lighting, alarm systems and industrial equipment applications (EPA 1996). Although the EPA puts the recycling rate at 96 per cent for SLAB in the United States, it has become clear that much of this activity is taking place offshore. From 2000 to 2007, US exports of SLAB and other lead waste increased by more than 900 per cent, with more than 1.2 million tons of SLAB and lead waste exports in 2006. This chapter examines the composition of US SLAB and lead waste exports across countries and the links between country characteristics that are associated with these large and changing export flows. In particular, the role of differences in the stringency of environmental standards for explaining SLAB and lead waste exports is examined. The data show that the composition of US exports to developing countries has increased dramatically in the past decade and that these countries typically do have lower environmental stringency. However, these countries also have many other characteristics that are important for trade that turn out to be potentially more influential factors than environmental stringency alone. In particular, the analysis to follow shows that income, organized crime and hidden trade barriers all play important roles in the determination of the location of US exports of SLAB and lead waste.

Environmental regulation across countries and the composition of US SLAB and lead waste exports

To examine the question of whether differences in environmental stringency across countries are a contributing factor for the location of US exports of SLAB and lead waste, two primary datasets were assembled. The first was data on two categories of waste exported by the United States and comes from the United Nations COMTRADE Database:[2] HS code 854810 (Waste & scrap of primary cells and batteries) and HS code 780200 (Lead waste or scrap). The second primary source of data comes from the 2000–2007 Global Competitiveness Reports (GCR) that are produced by the World Economic Forum. The GCR provide annual survey results of executives across countries and industries regarding a wide variety of aspects that affect business in those countries. One of the questions asked in the survey

Table 7.1 World Competitiveness Report survey questions

Variable	*Question*
Stringency of environmental regulations	How stringent are your country's environmental regulations? (1 = lax compared with those of most countries, 7 = among the world's most stringent)
Hidden trade barriers	In your country, hidden import barriers (that is, barriers other than published tariffs and quotas) are (1 = an important problem, 7 = not an important problem)
Organized crime	Organized crime (e.g. mafia-oriented racketeering, extortion) in your country (1 = imposes significant costs on business, 7 = does not impose significant costs on business)

has to do with the relative stringency of environmental regulations in a country.[3] The exact question asked regarding environmental stringency can be found in the first row of Table 7.1. Due to limited country coverage prior to 2000 and discontinuation of the environmental stringency question in the survey beyond 2007, the dataset employed in this chapter is constrained to the years 2000–2007, the years for which data is available for a wide variety of countries to which the United States exports SLAB and lead waste. Fortunately, this happens to be an intriguing period over which there was a large transformation in the quantity and composition of US exports of SLAB and lead waste.

In Table 7.2 we see that the total quantity of SLAB and lead waste exported from the United States between 2000 and 2007 increased dramatically, with the peak being 1.2 million tons in 2006. What is more dramatic than the rapid rise in the total tonnage of SLAB and lead waste exports over such a short time period was the change in the composition of exports. In 2000, developing countries only imported 17 per cent of US SLAB and lead waste. By 2007, developing countries were importing nearly 85 per cent of the US SLAB and lead waste exports. Over this very short time period, not only was there a drastic increase in the total volume of waste shipped from the United States to other countries, but the composition of waste going to developing countries increased in dramatic fashion.

The primary question in this study is whether differences in environmental stringency played a significant role in this change in US SLAB and lead waste exports. In Table 7.2, we see the average environmental stringency score for developed and developing countries that received imports from the United States. Throughout the 2000–2007 period, the developing country scores are significantly lower, indicating a lower level of environmental stringency. However, if changes in relative levels of environmental stringency across countries are a driving force for the shipments of SLAB

Table 7.2 US SLAB and lead waste exports and importing country environmental stringency

Year	*US SLAB and lead waste exports (tons)*[a]				*Average importing country environmental stringency index*[b]		
	Total US SLAB and lead waste exports	*To developing countries*	*To developed countries*	*To developing countries as % of total*	*Developing countries*	*Developed countries*	*Developing countries as % of developed countries*
2000	85,397	14,563	70,833	17.05	3.02	5.36	0.56
2001	127,113	67,056	60,057	52.75	3.18	5.60	0.57
2002	117,579	67,101	50,478	57.07	3.27	5.40	0.61
2003	102,237	52,366	49,871	51.22	3.33	5.49	0.61
2004	379,393	323,550	55,842	85.28	3.42	5.50	0.62
2005	717,251	654,261	62,990	91.22	3.42	5.58	0.61
2006	1,215,162	1,096,595	118,567	90.24	3.61	5.63	0.64
2007	774,622	654,233	120,388	84.46	3.57	5.35	0.67

Notes: [a] Data is for HS categories 854810 and 780200 from the UN COMTRADE Database.

[b] Data is from the 2000–2007 Global Competitiveness Reports.

Table 7.3 Top ten importers of US SLAB and lead waste from 2000 to 2007

Country	*SLAB and lead waste imported from United States (tons)*	*% of total US SLAB and lead waste exports*
Mexico	2,724,716	77.55%
Canada	469,200	13.35%
China	158,906	4.52%
Korea	74,318	2.12%
India	31,543	0.90%
Spain	14,434	0.41%
France	6,325	0.18%
Hong Kong	4,759	0.14%
Dominican Republic	4,232	0.12%
United Kingdom	3,620	0.10%

and lead waste, then Table 7.2 holds little preliminary evidence to support such large changes in the US exports to developing countries. Although the average developing country environmental stringency score was much lower than the average developed country score over the time period, the average score for developing countries was actually improving relative to the average score of developed countries. In 2000, the average developing country environmental stringency score was only 57 per cent of the average developed country score. By 2007, the average developing country environmental stringency score had improved to 67 per cent of the average developed country score. Thus, although the average developing country did indeed have lower environmental regulation stringency than the averaged developed country, developing country stringency was improving relative to the developed countries over the 2000–2007 time period.

Of course, US exports of SLAB and lead waste were not uniform across importers. During this time period, the United States exported SLAB and lead waste to 68 different countries, but exports were dominated by a few large importers, with Mexico being the largest importer by a substantial margin. Table 7.3 reports the top ten importers of SLAB and lead waste from the United States between 2000 and 2007. Given their close proximity to the United States, it is not a surprise that Mexico and Canada are the two largest importers, but vast differences in the two countries' volumes of SLAB and lead waste imports from the United States is striking.

Canada's average environmental stringency score for the period was much higher than Mexico's, with an average score over the time period of 5.8 compared to Mexico's 3.8. However, it is notable that Mexico's environmental regulation stringency score was improving over the time period, from 3.5 in 2000 to 3.9 in 2007, while Canada's environmental regulations stringency score was declining, from 5.9 in 2000 to 5.4 in 2007. Further, Mexico's environmental stringency score was actually higher than the average developing country score. Thus, three things become apparent from the data: (1) although Mexico had a lower environmental stringency score than Canada, its environmental stringency was improving relative to Canada over the time period; (2) despite a rapid increase in Mexico's imports of SLAB and lead waste from the United States, from 3,699 tons in 2000 to over 560,000 tons in 2007, Mexico was actually improving its environmental regulation stringency score; and (3) Mexico's average environmental score was higher than the average developing country score. On the surface, these three pieces of information suggest that environmental stringency, or at least environmental stringency alone, may not be the only factor driving changes in Mexican imports of US SLAB and lead waste.

However, Mexico's dominant volumes over Canada as a border country and the fact that two other developing countries, China and India, make up two of the next three spots in the top ten raises questions about what characteristics might be important in driving these trade flows. Are environmental regulation differences a critical factor or are other socio-economic, geographic or industrial factors important?

Data and model

To test the question of the role of environmental regulation stringency for US exports of SLAB and lead waste, the UN COMTRADE data on US exports and the WCR data on environmental stringency were combined with additional data to estimate a panel version of the PPML gravity model estimator developed by Santos Silva and Tenreyro (2006). Given the panel nature of the dataset, country importer fixed effects are employed to control for all unobserved country heterogeneity that is unchanged over time. As such, traditional geographic differences such as common language, distance, border dummies, cultural or religious differences, or landlocked and island dummies are captured by the importing country fixed effects. What remains then is to account for other important gravity model or institutional variables that may affect imports of US SLAB and lead waste but which do not remain constant over time.

These additional explanatory variables employed in the estimation are GDP, GNI per capita, capital stock per labourer, a dummy variable for whether the importing country has a free trade agreement with the United States, and two additional survey questions from the WCR regarding hidden trade barriers and organized crime in the importing country.[4] The questions regarding organized crime and hidden trade barriers can be found in Table 7.1. They are included in the model because Clark *et al.* (2004) show that organized crime can decrease port services and increase transport costs for trade in general. Further, there is evidence to suggest that organized crime groups often are involved in waste industry businesses (Tsai 2008; Interpol

2009; Lamendola 2011) and could have influences on trade in these particular types of industries. To the extent this is true, countries with high levels of organized crime will be more likely to import waste products. Likewise, hidden trade barriers, which can include anything from bribes to customs inspections to unobserved transportation challenges, may also play a role in determining trade flows to a country. All else equal, we expect that if an importing country has high levels of hidden trade barriers, then there will be less trade to those countries. To ease in variable interpretation, the WCR scores, which range from one to seven, were transformed such that a high score now corresponds to a higher degree of hidden trade barriers or costs from organized crime and a low score corresponds to a low degree of hidden trade costs or costs from organized crime.[5]

One potentially important policy variable is whether an importing country is a ratified member of the Basel Convention on the Trans-boundary Movements of Hazardous Waste and Their Disposal. However, none of the 68 countries in the sample changed their ratification status during the time period analysed in this study. Thus, any effects of ratification of the Basel Convention are captured by the importing country fixed effects in the regression model.

Table 7.4 provides descriptive statistics for all observations in the sample and also breaks the sample down by country development group (developed and developing). One thing that is apparent is that the developing countries and developed countries have significant differences in their average characteristics in the observables. First, the average developing country imports more than 3.5 times as much SLAB and lead waste from the United States than the average developed country. Mexico certainly has much to do with this fact. However, if we remove the two border countries (Mexico and Canada) from the dataset, developing countries still import a greater average volume of SLAB and lead waste than developed countries, with a developing country average of 334 tons to 269 tons for developed countries. Developed countries also have larger economies, higher incomes, a smaller impact from organized crime, fewer hidden trade barriers, and substantially more capital per worker than developing countries. Importantly, the average developing country in the sample has substantially lower environmental regulation stringency than the average developed country, 3.4 in developing countries compared to 5.5 in developed countries.

As mentioned above, the primary goal in this study is to attempt to isolate the role of environmental regulation as a predictor of US SLAB and lead waste exports, while controlling for other important gravity model factors that have been important in the literature. The gravity model is structured such that we have observations for i importers, over j waste categories (SLAB waste and lead waste), in year t. The PPML estimator for waste trade between the United States and the importing countries has a conditional mean, μ, that depends on time varying country characteristics $\mathbf{x}_{it}$, a vector

Table 7.4 Descriptive statistics for all countries and by development group

	All countries				
Variable	*Obs*	*Mean*	*Std. Dev.*	*Min*	*Max*
Tons of SLAB and lead waste exported	1,012	3,472	45,810	0	1,086,449
GDP (billion 2000 US $)	1,012	360.1	719.0	3.7	5,200.0
GNI per capita (thousand 2000 US $)	1,012	11.43	11.78	0.28	43.50
Capital stock per labourer (thousand 2005 PPP $)	1,012	79.26	64.80	2.08	240.56
Free trade area	1,012	0.09	0.28	0	1
Environmental stringency	1,012	4.3	1.3	1.9	6.8
Hidden trade barriers	1,012	2.3	1.0	0.2	4.5
Organized crime	1,012	2.2	1.3	0.1	5.3
	Developing countries				
Tons of SLAB and lead waste imported	624	4,692	57,931	0	1,086,449
GDP (billion 2000 US $)	624	160.4	315.3	3.2	2,500.0
GNI per capita (thousand 2000 US $)	624	2.32	1.90	0.25	9.09
Capital stock per labourer (thousand 2005 PPP $)	624	26.06	15.09	1.86	70.47
Free trade area	624	0.08	0.27	0	1
Environmental stringency	624	3.4	0.7	1.9	5.3
Hidden trade barriers	624	3.0	0.6	0.8	4.5
Organized crime	624	3.0	1.0	0.6	5.3
	Developed countries				
Tons of SLAB and lead waste imported	448	1,307	8,000	0	81,651
GDP (billion 2000 US $)	448	591.8	968.0	8.7	5,200.0
GNI per capita (thousand 2000 US $)	448	22.61	9.17	4.32	43.50

Table 7.4 (cont.)

		All countries			
Variable	*Obs*	*Mean*	*Std. Dev.*	*Min*	*Max*
Capital stock per labourer (thousand 2005 PPP $)	448	143.23	42.19	45.87	240.56
Free trade area	448	0.08	0.28	0	1
Environmental stringency	448	5.5	0.8	3.3	6.8
Hidden trade barriers	448	1.4	0.7	0.2	3.4
Organized crime	448	1.2	0.8	0.1	4.0

of coefficients β, importing country dummy variables d_i^I, year dummy variables d_i^I, and is given by:[6]

$$\mu_{it} = E_{it}(W_{it} \mid X_{it}) = \exp\left[\left(\prod_{i=1}^{N} d_i^I \prod_{t=1}^{T} d_t^T\right) X_{it}\beta\right] \tag{1}$$

Parameter estimates are obtained by maximizing the log of the likelihood function,

$$L(\beta W, X) = \prod_{i=1}^{N}\prod_{t=1}^{T} \Pr(W_{it} \mid \mu_{it}) = \prod_{i=1}^{N}\prod_{t=1}^{T} \frac{\exp(-\mu_{it})\mu_{it}^{W_{it}}}{W_{it}!} \tag{2}$$

with errors clustered on the country-year observation.

Results and discussion

One of the primary concerns in interpreting parameter estimates for the environmental stringency variable from Equation (2) is the fact that environmental stringency tends to be highly correlated with other important explanatory variables. Those countries that have higher incomes also tend to have higher capital/labour ratios and subsequently are able to afford and demand higher environmental regulations. Indeed, many papers use income as a proxy for the stringency of environmental quality across countries (Antweiler *et al.* 2001 or Baggs 2010). In Table 7.5, we see that environmental stringency has a strong positive correlation with GNI per capita and the capital/labour ratio, but that it also has a strong negative correlation with hidden trade barriers and organized crime. Countries with high incomes and strong environmental stringency also tend to have lower hidden trade barriers and lesser costs associated with organized crime.

Table 7.5 Simple correlations between environmental stringency and covariates

Variable	*Environmental stringency*
Environmental stringency	1.00
GDP	0.33
Free trade area	0.05
GNI per capita	0.85
Capital/labour ratio	0.81
Hidden trade barriers	−0.78
Organized crime	−0.72

Recognizing the strong correlations amongst environmental stringency and several of the other time-variant explanatory variables, the model is first estimated in column (1) of Table 7.6 with only GDP and the Free Trade Area dummy, the two variables that are not correlated with environmental stringency in Table 7.5, included with the environmental stringency variable. The positive and statistically significant environmental stringency variable indicates that, all else being equal, more stringent environmental regulations lead to more imports of SLAB and lead waste from the United States for an importing country. This is counterintuitive to the hypothesis that lower environmental regulation stringency leads to more waste imports for countries. However, we must keep in mind that this parameter estimate may be picking up a number of the other characteristics that we have left out of the regression that are simply correlated with environmental stringency levels.

In columns (2) through (5) of Table 7.6 each of the variables that are strongly correlated with environmental stringency are added to the regression. The addition of GNI/capita decreases the magnitude of the environmental stringency variable, with GNI/capita having a strong negative impact on imports of US SLAB and lead waste. The addition of the capital/labour ratio changes the parameter estimates slightly but the coefficient is not significant and has no impact on the qualitative results of the model. In columns (4) and (5), the environmental stringency variable is no longer statistically significant when hidden trade barriers and organized crime are added to the model.

The progression from column (1) to column (5) highlights some important characteristics regarding the influences on imports of US SLAB and lead waste. First, the stringency of environmental regulations does not appear to be the driving factor in determining the location of US SLAB and lead waste flows when other factors that tend to be correlated with environmental stringency are taken into account. US SLAB and lead waste tends to be influenced by small economies with lower levels of income, fewer hidden trade barriers, and higher levels of organized crime. To test the robustness of these results, environmental stringency is dropped from the regression

Table 7.6 PPML regressions on US SLAB and lead waste exports

	(1)	*(2)*	*(3)*	*(4)*	*(5)*	*(6)*
Environmental stringency	1.710***	1.067*	1.293**	0.317	0.247	
	(0.639)	(0.558)	(0.648)	(0.734)	(0.734)	
GDP	−0.005***	−0.005***	−0.005***	−0.006***	−0.007***	−0.007***
	(0.002)	(0.001)	(0.001)	(0.002)	(0.002)	(0.002)
Free trade area	−0.508	−0.191	−0.11	−0.406	−0.654	−0.692
	(0.544)	(0.598)	(0.637)	(0.677)	(0.748)	(0.725)
GNI per capita		−1.101***	−1.405*	−1.457*	−1.617*	−1.580*
		(0.357)	(0.814)	(0.820)	(0.847)	(0.827)
Capital/labour ratio			0.032	0.030	0.041	0.033
			(0.062)	(0.060)	(0.062)	(0.055)
Hidden trade barriers				−1.060**	−1.358***	−1.444***
				(0.443)	(0.456)	(0.389)
Organized crime					0.954*	0.960*
					(0.580)	(0.580)
Log psuedoliklihood	−2,205,614	−2,129,519	−2,127,135	−2,095,840	−2,080,366	−2,080,974
Observations	1,072	1,072	1,072	1,072	1,072	1,072

Notes: Standard errors, clustered on country–year observations, in parentheses. Year dummies and importer fixed effects are included in all regressions.
* significant at 10%; ** significant at 5%; *** significant at 1%.

Table 7.7 PPML robustness regressions on US SLAB and lead waste exports

	(1)	*(2)*	*(3)*	*(4)*
	All countries (Mexico excluded)	*Developing*	*Developing (Mexico excluded)*	*Developed*
Environmental stringency	1.303**	−0.7	2.722***	−0.728
	(0.523)	(1.611)	(0.419)	(0.601)
GDP	−0.003***	−0.008**	−0.002**	0.003
	(0.001)	(0.003)	(0.001)	(0.004)
Free trade area	−0.541	−2.351***	−0.288	0.055
	(0.558)	(0.828)	(0.534)	(0.757)
GNI per capita	0.069	0.04	−1.289*	0.386***
	(0.152)	(1.661)	(0.735)	(0.146)
Capital/labour ratio	0.013	−0.802*	0.22	−0.003
	(0.024)	(0.441)	(0.149)	(0.031)
Hidden trade barriers	−0.431*	−1.822***	−0.752***	−1.053*
	(0.232)	(0.284)	(0.192)	(0.563)
Organized crime	0.780*	2.873***	0.724***	0.048
	(0.441)	(0.623)	(0.267)	(0.369)
Log psuedoliklihood	−207,734	−829,011	−42,854	−119,104
Observations	1,056	624	608	448

Notes: Standard errors, clustered on country–year observations, in parentheses. Year dummies and importer fixed effects are included in all regressions. * significant at 10%; ** significant at 5%; *** significant at 1%.

in column (6). The coefficient estimates on the other explanatory variables remain remarkably similar to those in column (5).

To further test the robustness of the results and to see if there are differences in the results for developing and developed countries, additional estimates are presented in Table 7.7. In column (1), the model is run on the full sample but with Mexico dropped. Given Mexico's prominent role in imports of US SLAB and lead waste it is not surprising that there are some changes in the magnitudes of the coefficients in the models. Qualitatively, however, the results are fairly similar to those in column (5) of Table 7.6. US SLAB and lead waste exports tend to be influenced by smaller economies with fewer hidden trade barriers and more organized crime. The one substantial difference is that environmental stringency is now positive and significant. One interpretation of this is that the importance of environmental regulation stringency in Mexico, the largest importer of US SLAB and lead waste, is quite different than for other importing countries. All else equal,

US SLAB and lead waste exports may be attracted to places with higher environmental quality, with Mexico being a huge and significant exception which would bias the environmental stringency coefficient down in column (5) of Table 7.6 relative to the coefficient in column (1) of Table 7.7.

In column (2) of Table 7.7, the model is run just on the developing country observations. The results are similar to the results for the overall sample in column (5) of Table 7.6, although the magnitude of the hidden trade barriers and organized crime coefficients are substantially larger, indicating that hidden trade barriers and organized crime play a more prominent role for developing countries. In column (3), Mexico is dropped from the developing country sample. Interestingly, when Mexico is dropped, the coefficients for hidden trade barriers and organized crime fall by an order of magnitude and the coefficient on environmental stringency is now positive and significant. Clearly, the inclusion of Mexico tends to bias the coefficients for environmental stringency and hidden trade barriers downwards and the coefficient for organized crime upwards. In column (4), environmental stringency and organized crime are statistically insignificant factors for developed countries, although hidden trade barriers remain negative and statistically significant.

The results of the regressions in Table 7.6 and Table 7.7 provide interesting insights into the factors that influence US SLAB and lead waste exports. Overall, the presence of hidden trade barriers tends to be the most consistent statistically significant factor in discouraging imports into a country. Smaller economies and the prevalence of organized crime in a country also tend to be strong predictors of imports of US SLAB and lead waste, although organized crime and country size largely seem to be driven by the experience of the developing countries rather than the developed countries.

Conclusions

The increased integration of markets across the globe is a trend that does not seem likely to show any signs of slowing in the coming decades. A great deal of attention has been placed on studying the effects of these transformations in high-profile markets such as the financial sector and manufacturing. However, growing market integration and trade in less glamorous industries, such as waste industries, have been expanding at a rapid rate, with far less attention paid to the influences and consequences of these trade flows. These transformations not only have economic and financial impacts on many industries but also have potentially large environmental consequences for many countries.

This chapter examines the US SLAB and lead waste industries and their exports over an eight-year period for which there was exceptional growth. With an average annual growth rate of 60 per cent, US exports of SLAB and lead waste grew from 85,397 tons in 2000 to 774,622 tons in 2007, which included a dramatic shift in the proportion of those exports going to developing countries. The results of this chapter suggest mixed evidence that environmental regulation stringency was a statistically significant influence on these changes. All else equal, greater environmental regulation stringency had a positive impact on attracting US SLAB and lead waste imports. However, this result disappears when Mexico is included, suggesting that the dominant experience of Mexico may be an important exception to that result. Other factors such as the size of the economy, the prevalence of hidden trade barriers and organized crime are shown to be more robust and important factors in explaining the location of US exports of SLAB and lead waste, particularly in developing countries.

Despite the mixed results for the environmental regulation stringency variable in this chapter, it is important to point out that environmental regulation stringency has a high negative correlation with organized crime across countries, which is a statistically significant and meaningful factor in this analysis. This is very important for two reasons. First, developing countries, who have seen the largest increases in imports of SLAB and lead waste from the United States, have more organized crime and lower environmental regulation stringency than developed countries. Whether it is organized crime or low environmental regulation stringency that is the statistically significant factor is somewhat irrelevant from an environmental standpoint. It implies that more SLAB and lead waste will end up in these developing countries. Without adequate environmental regulations, this can lead to environmental consequences for the importing country. This chapter does not address the environmental consequences of these trade flows, but this remains an important area of future research.

Second, the statistical significance of organized crime and its correlation with lower environmental regulation stringency underscores an important international policy point when addressing trade in waste products such as SLAB and lead waste. Efforts to reduce waste trade flows with more stringent environmental policy across countries may have little influence if there is not also a directed effort at controlling organized crime.

Notes

1 See for example, Ederington and Minier (2003) for the effects of environmental regulation on trade or Ederington *et al.* (2005) or Kellenberg (2009b) for the effects of environmental regulation on FDI.
2 The database can be found at http://comtrade.un.org/db/.
3 This measure of environmental stringency has also be used in Kellenberg (2009b) and Kellenberg (2012).
4 Data on GDP, GNI per capita and labour force were obtained from the World Bank Development Indicators database. Capital stock was calculated using the perpetual inventory method and capital investment data from the Penn World Tables 6.3. Data on free trade agreements was obtained from the World Trade Organizations statistics database at www.wto.org.
5 The exact transformation was to take 7 – (WCR score) = new score. For example, a country with an organized crime score of 2.1 would have a transformed score of 7 – 2.1 = 4.9.

6 Since $j = 2$ in this case, to simplify notation a dummy variable is included in $\mathbf{x}_{it}$ to indicate whether the observation is lead waste rather than SLAB and the j subscript is dropped from Equations (1) and (2).

References

Antweiler, W., Copeland, B.R. and Taylor, M.S. (2001) 'Is Free Trade Good for the Environment?', *American Economic Review*, 91(4): 877–908.

Baggs, J. (2009) 'International Trade in Hazardous Waste', *Review of International Economics*, 17(1): 1–16.

Clark, X., Dollar, D. and Micco, A. (2004) 'Port Efficiency, Maritime Transport Costs, and Bilateral Trade', *Journal of Development Economics*, 75: 417–50.

Ederington, J. and Minier, J. (2003) 'Is Environmental Policy a Secondary Trade Barrier? An Empirical Analysis', *Canadian Journal of Economics*, 36(1): 137–54.

Ederington, J., Levinson, A. and Minier, J. (2005) 'Footloose and Pollution-Free', *The Review of Economics and Statistics*, 87(1): 92–9.

EPA (1996) *Implementation of the Mercury-Containing and Rechargeable Battery Management Act*, Washington, DC: EPA, www.epa.gov/epawaste/hazard/recycling/battery.pdf.

Interpol (2009) 'Electronic Waste and Organized Crime: Assessing the Links', *Trends in Organized Crime*, 12: 352–78.

Kellenberg, D. (2009a) 'U.S. Affiliates, Infrastructure, and Growth: A Simultaneous Investigation of Critical Mass', *The Journal of International Trade and Economic Development*, 18(3): 311–45.

Kellenberg, D. (2009b) 'An Empirical Investigation of the Pollution Haven Effect with Strategic Environment and Trade Policy', *Journal of International Economics*, 78(2): 242–55.

Kellenberg, D. (2012) 'Trading Wastes', *Journal of Environmental Economics and Management*, 64(1): 68–87.

Lamendola, M. (2011) *Report Says Organized Crime Still Prevalent in Waste Management*, www.northjersey.com/news/135638183_Report__Mob_still_tied_to_waste_management.html?page=all.

Santos Silva, J.M.C. and Tenreyro, S. (2006) 'The Log of Gravity', *The Review of Economics and Statistics*, 88(4): 641–58.

Tsai, M. (2008) 'Why the Mafia Loves Garbage: Hauling Trash and Organized Crime', *Slate*, www.slate.com/articles/news_and_politics/explainer/2008/01/why_the_mafia_loves_garbage.html.

8 The political cost of residual municipal solid waste taxation

Perception versus reality

Simon De Jaeger

Introduction

This chapter makes a contribution to the political economy literature by analysing whether and how consumers hold municipal incumbents responsible for residual municipal solid waste (MSW) pricing policies. The political economy literature proposes several theories in relation to the information taken into account by the electorate when judging incumbents' performance. We model and test empirically the three most common scenarios. In the first scenario the basic consumer choice problem is based on the models introduced in the seminal works of Kinnaman and Fullerton (2000) and Fullerton and Kinnaman (1995). The representative consumer maximizes its utility subject to its waste balance equation and its budget constraint, but ignores the budget constraint faced by local policymakers when deciding residual MSW prices. In this setting, consumers' utility decreases as residual MSW prices increase. Based on the median voter literature (Downs 1957), the second scenario assumes that consumers concede that municipalities need to finance the cost of waste processing and disposal services, but also understand that lump sum transfers within the municipality can play a role. The resulting model specification shows that utility maximizing residual MSW prices depend on the consumer's relative level of MSW generation. If the consumer produces less (more) residual MSW than the average residual MSW production in his municipality, utility increases (decreases) as the price increases. In the third scenario we assume that consumers compare the MSW price in their municipality with the price charged in neighbouring municipalities, and use the latter as a yardstick to measure incumbents' performance. Consumers (voters) will adopt this method to judge whether the residual MSW price in their municipality is acceptable, if they lack information (Besley and Case 1995).

We test empirically the predictions from each scenario using prices for residual MSW collection and disposal services and popularity scores for the 308 Flemish municipalities, in 2006 and 2009. The test results allow us to determine which of the three scenarios is supported by the data. This is relevant since there are numerous empirical studies in the field of strategic income and property tax policy interaction among local governments that show the existence of spatial policy interaction (Ashworth *et al.* 2006; Ashworth and Heyndels 1997; Bosch and Solé-Ollé 2007; Solé-Ollé 2003; Revelli 2001, 2002; Bordignon *et al.* 2003, 2004; Geys and Vermeir 2008; Brueckner 2003; Wilson 1996). For Belgium Heyndels and Vuchelen (1998) show that local income and property tax policies are strongly correlated among neighbouring municipalities, and De Jaeger *et al.* (2009) find that local jurisdictions also strategically interact with each other when deciding on the waste price. A municipality counteracts a fall of 1 in the waste price applied by neighbouring municipalities by decreasing its price by 0.23. Testing the predictions from our model in the third scenario should enable us to confirm whether such residual MSW price mimicking is advantageous in relation to re-election chances.

The impact of tax variables on the popularity scores of incumbents and the vote shares of policymakers in general, has received substantial attention.[1] Although most authors focus on nationwide or state data, there is also empirical evidence at the local (e.g. municipal) level. Several papers use the concept of yardstick voting to empirically test the political costs of taxes at the sub-national level. For instance, the results in Bosch and Solé-Ollé (2007) suggest that an increase in property taxes had an impact on the incumbent's share of the vote in Spanish municipalities. However, the political costs of an increase in property taxes seem to be conditional on the tax policy in neighbouring municipalities. To our knowledge, the only empirical analysis on Flemish data is Vermeir and Heyndels (2006) who analyse municipal elections in Flanders over the period 1982–2000 and find that the cost in terms of re-election chances depends on neighbouring municipalities' local income tax and property tax rates (the two categories that generate the most revenue). The impact of residual MSW prices on incumbents' popularity rates and vote shares has not previously been analysed.

It could be argued that consumers are indifferent about waste prices since the cost of residual solid waste collection and processing services represents only a marginal fraction of total household expenses (based on the municipal accounts and income statistics we estimate that, in Flanders, only 0.17 per cent of average income in 2006 went on waste disposal and processing costs). However, in Flanders, and in many other regions and countries, unit based pricing is employed and residents have to buy special waste bags at local shops for kerbside collection. This 'reminds' residents of the price each time they have to buy bags. Also, it makes it easier to compare these prices across municipalities, which might induce yardstick competition. We believe, therefore, that prices for residual MSW might have an impact on incumbents' popularity scores and vote shares.

Next section presents our model of consumers' preferences regarding local municipalities' residual MSW pricing policies under the three scenarios discussed. Following sections discuss the data and methodology and present the empirical estimates. Final section concludes.

The model

We start with a basic consumption choice model for a representative consumer in a given municipality i. The consumer observes the price for residual MSW disposal services (henceforth waste price) set by the local municipality and decides about the levels of his consumption and waste generation. The representative consumer then uses the waste price to assess the relative performance of local decision makers. However, we do not know what other information is taken into account by the representative consumer when judging the pricing policy of the incumbent. Therefore, we explore three scenarios in which the consumer's level of information differs. In the first scenario, the model set-up of the basic consumers' choice problem is based on the models introduced by Kinnaman and Fullerton (2000) and Fullerton and Kinnaman (1995). The consumer considers only the waste prices in municipality i. When judging the incumbent, the consumer considers his/her utility which he/she maximizes with respect to their particular budget constraints. The second scenario draws on the median voter literature (Downs 1957 for technical details on the median voter model, Congleton 2003 for an introductory review). We build also on the previous model, but assume that consumers also take account of local policy makers' budget constraints. Consumers realize that municipalities have to finance the costs of waste disposal services, but also understand the role of lump sum transfers within municipalities. In the third scenario, we introduce the concept of yardstick voting (Besley and Case 1995). The price of waste disposal may differ from the one preferred by the representative consumer, but the consumer assesses the incumbent based on the difference in utility in the event the price is equal to a certain benchmark. Since consumers are likely to have imperfect information we assume they use as a benchmark either the price in a neighbouring municipality or historic prices in their own.

Scenario 1

The set-up of the basic consumers' choice problem mimics the models proposed by Kinnaman and Fullerton (2000) and Fullerton and Kinnaman (1995). We assume a consumer who derives utility from consuming a composite consumption good x, but also experiences disutility from the total volume of waste in his municipality (W).[2] For example, there is evidence suggesting that waste processing facilities have a negative impact on the value of real estate (Kiel and McClain 1995). Also, the external costs associated with the traffic caused by waste collection may increase disutility (for an extensive overview of potential external costs related to landfilling and incineration of solid waste see Eshet *et al.* 2006). We assume also that consumption of x gives rise to an amount $w = \alpha(e)\cdot x$ of solid waste. The parameter $\alpha > 0$ is the average waste-to-consumption ratio and depends on the effort level $e \in [0,1]$ provided by the representative consumer to reduce his/her waste-to-consumption ratio.

Since it becomes increasingly difficult to continuously reduce the average waste-to-consumption ratio we assume $\alpha' < 0$ and $\alpha'' > 0$. Reducing α is costly and time consuming. It takes time to engage in at home recycling or to find a substitute consumption good that generates less waste. We assume, therefore, that the consumer incurs a cost to provide this effort and that this cost is increasing in the amount of effort provided: $c(e)$ with $c(0)=0, c' \geq 0, c' \geq 0, c'(0) = 0$, and $\lim_{z\to 1} c'(z) = +\infty$. Consumers have a limited budget $\bar{y}$ that they can use to finance consumption and resulting waste disposal or waste reduction activities. The representative consumer maximizes his/her utility subject to her waste balance equation and budget constraint. Using subscripts to indicate the scenario (1, 2 or 3) we can represent the consumers' choice problem as:

$$\max_{x_1, w_1, e_1} u_1(x_1, W_1)$$
$$s.t. \begin{cases} \alpha_1(e_1)\bullet x_1 = w_1 \\ 1\bullet x_1 + p_w \bullet w_1 + c_1(e_1) \leq \bar{y} \end{cases} \tag{1}$$

where the price of the composite consumption good is normalized to one (*numéraire* good). p_w is the price for disposing of one unit of waste in the regular way (i.e. presenting it at the kerbside in the mandatory container, e.g. bag or bin). The Lagrange function for this consumer problem is given by:

$$L_1(x_1, e_1, \lambda_1) = u_1(x_1, \bar{W}_1) - \lambda[1\bullet x_1 p_w \bullet(\alpha_1(e_1)\bullet x_1) + c_1(e_1) - \bar{y}] \tag{2}$$

Necessary first-order conditions for a solution to this problem are given by:[3]

$$\begin{aligned} \frac{\partial L_1}{\partial x_1} &= \frac{\partial u_1}{\partial x_1} - \lambda_1 \bullet 1 - \lambda_1 \bullet \alpha_1(e_1)\bullet p_w = 0 \Rightarrow \frac{1}{\lambda_1}\bullet\frac{\partial u_1}{\partial x_1} = 1 + \alpha_1(e_1)\bullet p_w \\ \frac{\partial L_1}{\partial e_1} &= -\lambda_1 \bullet p_w \bullet \frac{\partial \alpha_1}{\partial \varepsilon_1}\bullet x_1 - \lambda_1 \bullet \frac{\partial c_1}{\partial e_1} = 0 \Rightarrow \frac{\partial c_1}{\partial e_1} = -p_w \bullet \frac{\partial \alpha_1}{\partial e_1}\bullet x_1 \end{aligned} \tag{3}$$

The first condition says that marginal utility (in monetary terms) of consuming an additional unit of the composite good should be equal to the price of the unit (normalized to one) plus the cost of disposing its waste ($\alpha_1(e_1)\bullet p_w$). The second condition states that marginal benefit of effort equals the disposal cost saved.

Combining conditions (3) and the total derivatives of the constraints, we can show that demand for the consumption good (x_1) decreases and effort (e_1) increases in the price for waste disposal services (p_w). Note effort $(e_1(p_w, \bar{y}))$, demand for the consumption good $(x_1(p_w, \bar{y}))$ and waste ($w_1 = \alpha_1(w_1)\bullet(e_1)\bullet x_1$) are endogenously determined in our model.

$$dx_1 = -w_1 \cdot dp_w - p_w \cdot \alpha_1' \cdot de_1 \cdot - p_w \cdot \alpha_1 \cdot dx_1 - c_1'(e_1) \cdot de_1$$
$$\Rightarrow \frac{dx_1}{dp_w} \cdot (1 + p_w \cdot \alpha_1) = -w_1 - p_w \cdot \alpha_1' \cdot x_1 \cdot \frac{de_1}{dp_w} - c_1'(e_1) \cdot \frac{de_1}{dp_w} \tag{4}$$
$$\Rightarrow \frac{dx_1}{dp_w} = -\frac{w_1}{(1 + p_w \cdot \alpha_1)} < 0$$

and

$$c_1'' \cdot de_1 = -(\alpha_1' \cdot x_1 \cdot dp_w + p_w \cdot \alpha_1'' \cdot x_1 \cdot de_1 + p_w \cdot \alpha' \cdot dx_1)$$
$$\Rightarrow (c_1'' + p_w \cdot \alpha_1'' \cdot x_1) \cdot \frac{de_1}{dp_w} = -\alpha_1' \cdot x_1 - p_w \cdot \alpha_1' \cdot \frac{dx_1}{dp_w}$$
$$\Rightarrow (c_1'' + p_w \cdot \alpha_1'' \cdot x_1) \cdot \frac{de_1}{dp_w} = \frac{-\alpha_1' \cdot x_1 \cdot (1 + p_w \cdot \alpha_1)}{(1 + p_w \cdot \alpha_1)} + \frac{p_w \cdot \alpha_1'' \cdot \alpha_1 \cdot x_1}{(1 + p_w \cdot \alpha_1)} \tag{5}$$
$$\Rightarrow \overbrace{(c_1'' + p_w \cdot \alpha_1'' \cdot x)}^{+} \frac{de_1}{dp_w} = -\frac{\overbrace{(-\alpha_1' \cdot x_1)}^{+}}{(1 + p_w \cdot \alpha_1)}$$
$$\Rightarrow \frac{de_1}{dp_w} > 0$$

In order to integrate the above results into the incumbent's popularity functions, we follow the so-called responsibility hypothesis (Powell and Whitten 1993; Revelli 2002). As Vermeir and Heyndels (2006) argued the empirical literature on vote and popularity functions starts with the responsibility hypothesis, which states that voters vote for the party or politician from which they expect the highest utility gain. These expectations may depend on the politicians' electoral platform, and on the incumbents' past performance. On this rationale, we can state the popularity score of the incumbent (denoted V_1) will depend on the utility of the electorate ($V_1(u_1)$) with $V'_1(u_1) > 0$. Given the focus of this chapter, we are interested mainly in the impact of waste prices on the popularity score of the incumbent. However, differentiating utility $u_1(x_1, W_1)$ with respect to the waste price p_w and using the comparative statics of our model, shows du_1/dp_w cannot be signed:

$$
\begin{aligned}
du_1 &= \frac{\partial u_1}{\partial x_1} \bullet dx_1 + \frac{\partial u_1}{\partial W} \bullet dW \\
du_1 &= \frac{\partial u_1}{\partial x_1} \bullet \frac{\partial x_1}{dp_w} \bullet dp_w + \frac{\partial u_1}{\partial W} \sum_{j=1}^{n} \left(\alpha'_{1,j} \bullet e'_{1,j} \bullet x'_{1,j} + \alpha'_{1,j} \bullet \frac{\partial x_{1,j}}{\partial p_w} \right) \bullet dp_w \\
\frac{du_1}{dp_w} &= \left\{ \frac{dx_1}{dp_w} + \overbrace{\left[\sum_{j=1}^{n} \left(\overset{-}{\alpha'_{1,j}} \bullet \overset{+}{e'_{1,j}} \bullet \overset{+}{x'_{1,j}} + \overset{+}{\alpha'_{1,j}} \bullet \overset{-}{\frac{\partial x_{1,j}}{\partial p_w}} \right) \right] \bullet \overset{-}{\prod} 1i(x_{1,i}, W)}^{+} \right\} \bullet \overset{+}{\frac{\partial u_1}{\partial x_1}}
\end{aligned}
\tag{6}
$$

with $\prod_{1,i}(x_{1,i}W) \equiv \frac{\partial u_{1,i}}{\partial W} \Big/ \frac{\partial u_{1,i}}{\partial x_{1,i}}$. The change in utility depends on two opposite effects. First, as consumption of x_1 decreases in the waste price (see expression 4), dx_1/dp_w is negative in 6. Second, an increase in the waste price reduces overall waste production and the associated external costs. In other words, the model in scenario 1 predicts that the impact of the waste price on the popularity score depends on a consumption effect and the external costs of waste collection and processing.

Scenario 2

So far, we have not taken account of the policymaker's budget constraint; we have assumed implicitly that the representative consumer is concerned only about the waste balance equation and their personal budget constraint. However, it could be argued that consumers understand that municipalities must finance the cost of the services they provide (Tresch 2002). In particular, the total revenue from residual MSW pricing should be sufficient to cover the cost of waste disposal and processing services $(G = g \bullet W_2 = g \bullet \overline{w}_2 \bullet n)$ and finance a lump sum transfer (T) to each individual. The local government's budget constraint is then given by:

$$
p_w \bullet \overline{w}_2 \bullet n = g \bullet \overline{w}_2 n + T \bullet n \tag{7}
$$

where p_w, as before, is the price for disposing of one unit of waste, $\overline{w}$ is the average amount of waste generated per resident, n is the number of residents and g is the cost to the municipality of collecting and processing one unit of waste. Note that the transfer can be positive or negative depending on the amount of the total cost covered by the total revenue ($g \bullet \overline{w}_i \bullet n$ versus $p_w \bullet \overline{w} \bullet n$). A positive transfer can take the form of a reduction in some other municipal tax, while a negative transfer is likely to take the form of an increase in municipal taxes. Since in this scenario we assume that the consumer understand that municipality must finance its waste programme, the consumer's budget should account for the resulting transfers:

$$\begin{aligned}&1\bullet x_2 + p_w \bullet w_2 + c_2(e_2) - T \leq \bar{y}\\&\Rightarrow 1\bullet x_2 + p_w \bullet (\alpha_2 \bullet x_2) + c_2(e_2) - (p_w - g)\bullet \overline{w}_2 \leq\end{aligned} \tag{8}$$

Since consumers will not consider the impact on the lump sum transfer when deciding on their optimal levels of waste (W) and effort (e), condition 3 will still hold in scenario 2. However, the impact of the price on consumption of the composite good will be different:

$$\begin{aligned}&dx_2 = -w_2 \bullet dp_w - p_w \bullet \alpha_2' \bullet x_2 \bullet de_2 - p_w \bullet \alpha_2 \bullet dx_2 + (\overline{w}_2 + p_w \bullet \overline{w}_2' - g \bullet \overline{w}_2') \bullet dp_w - c_2' \bullet de_2\\&\frac{dx_2}{dp_w} = \frac{(\overline{w}_2 - w_2) + \overline{w}_2' \bullet (p_w - g)}{(1 + p_w \bullet \alpha_2)}\end{aligned} \tag{9}$$

From the above expression we can distinguish between two effects: the sign of dx_2/dp_w depends on the consumer's relative level of waste generation $(\overline{w}_2 - w_2)$ and the cost recovery rate $(p_w - g)$. In the first case $(\overline{w}_2 - w_2)$ if the consumer generates less waste than the average per consumer waste production $(\overline{w}_2)$ in his municipality $(\overline{w}_2 > w_2)$, the consumption of the composite good increases as the price for waste disposal facilities increases. In the second case, if the consumer produces more waste than the average produced per consumer in his municipality $(\overline{w}_2 < w_2)$, consumption of x_2 decreases as p_w increases. This result is intuitive because, for consumers with $\overline{w}_2 - w_2$, the additional costs for waste disposal due to an increase in the price, will be more than compensated for by the change in the lump sum transfer (if we ignore the cost recovery rate). On the other hand, for consumers with $\overline{w}_2 < w_2$, the additional cost of an increase of p_w, will not be covered by the change in the lump sum transfer. Note that if $\overline{w}_2 = w_2$ it follows that $dx_2/dp_w = 0$, or the changes in the cost are exactly compensated by the changes in the lump sum transfer. The second effect $(p_w - g)$ is driven by the fact that a change in the price has an impact on the average waste generation $(\overline{w}_2)$ and the corresponding total cost of waste disposal and processing services (G). If the variable waste price is insufficient to cover the cost of the waste disposal services, a reduction in average waste production will reduce the lump sum transfer from consumer to municipality and, hence, increase consumption of x_2. If the variable waste price more than compensates for the costs of the waste disposal services, a reduction in average waste production will reduce the transfer from municipality to consumer, thus reducing consumption of x_2.

Differentiating utility with respect to the waste price and using the comparative statics derived above allow us to determine the sign of the marginal utility in the second scenario:

$$\frac{du_2[x_2(p_w)]}{dp_w} = \left\{ \overset{+/-}{\frac{dx_2}{dp_w}} + \overbrace{\left[\sum_{j=1}^{n} \left(\overset{-}{\alpha'_{2,j}} \bullet \overset{+}{e'_{2,j}} \bullet \overset{+}{x_{2,j}} + \overset{-}{\alpha'_{2,j}} \bullet \overset{-}{\frac{\partial x_{2,j}}{\partial p_w}} \right) \right]}^{+} \bullet \prod 2i(x_{2,i}, W) \right\} \bullet \overset{+}{\frac{\partial u_2}{\partial x_w}} \tag{10}$$

The model in scenario 2 predicts that the impact of a change in the price on the incumbent's popularity score will depend on the consumer's relative level of waste generation, the cost recovery rate and the external costs associated with total waste generation.

Scenario 3

In the third scenario consumers compare utility using different benchmarks for the price (p_w^b). The price may differ from that preferred by the representative consumer. The utility loss due to this difference can be expressed as (Solé-Ollé 2003):

$$\pi = u_3'(x_3(p_w)) \bullet (x_3(p_w) - x_3(p_w^b)) \tag{11}$$

We assume they either use the price in neighbouring municipalities (scenario 3a) or historic prices in their own municipality as a benchmark (scenario 3b). Note that the consumer compares waste price levels, but ignores any differences in lump sum transfers. If the consumer uses the previous price in his/her own municipality, (p_w^b) for period t is simply the price in period t-1. In the yardstick model, the representative consumer uses prices and levels of garbage in a neighbouring municipality to assess the relative performance of local decision makers. In line with Equation (10), the consumer compares the utility change on the basis that he/she lives in the neighbouring municipality, where the charge for waste disposal services is (p_w^b). As in Solé-Ollé (2003) the utility loss due to the divergence between the real and benchmark policy is the indicator the consumer uses to evaluate the incumbent's performance. Therefore, in this scenario, popularity scores no longer depend on absolute utility levels (as in scenarios 1 and 2), but on the utility loss ($V_3(\pi)$) with $V'_3(\pi) < 0$. It is easy to show that scenarios 3a and 3b both predict that the popularity score will increase as the benchmark increases for a given own price, and decrease as the own price increases for a given benchmark.

Data

In Flanders (i.e. one of the three regions of Belgium), the 308 local municipalities and city councils organize the collection and disposal of household waste. They can decide independently about many of the practical details related to waste collection such as recycling schemes, information campaigns and, most importantly, financial contributions. In the following analysis we focus on the variable price for residual MSW disposal services (i.e. the weight or volume based price or waste price), defined as the average variable price (in eurocents) charged by the municipality to collect and process 1 litre of residual MSW.[4] We have converted the average price for 1 kg to the price per litre using the official OVAM conversion table to account for municipalities using weight based pricing schemes. The prices range between 0.4 and 7.4 eurocents per litre (see Table 8.1). On average 2009 prices are higher than 2006 prices.

To assess political costs we need information on incumbents' vote shares or popularity scores. Geys and Vermeir (2008) point out that the potential political cost of taxation consists of two, often strongly related, components. First, taxes might influence politicians' popularity. Second, taxes might jeopardize the probability of being elected for another term of office. In this chapter we focus on the first component because the highly fragmented political landscape and the existence of local parties at the municipal level make it difficult to compare vote shares among different elections. At the municipal level, political parties frequently split up or merge with another party, and some parties disappear or change their names between elections. Popularity ratings are available for 2006 and 2009 for all municipalities. The ratings are based on an online questionnaire administered by the online research bureau iVOX (in association with the Flemish newspaper *Het Nieuwsblad* and a radio station – 'Radio 2').[5] With more than 140,000 respondents, the sample comprises about 2.3 per cent of the total population in Flanders.[6] Each respondent was asked to rate his/her satisfaction with the relevant local authority on a Linkert scale. The average scores in Flanders (see Table 8.1) are similar for both reference years (approximately 6.7).

As mentioned in the introduction, empirical evidence in De Jaeger *et al.* (2009) shows that waste prices and waste quantities in Flanders exhibit spatial clustering, and the authors argue that waste prices depend endogenously on waste prices in nearby municipalities since local policymakers engage in tax mimicking. However, we have no reason to assume that popularity scores endogenously depend on the popularity scores in other, geographically close, municipalities (i.e. popularity scores have no direct effect on scores in neighbouring municipalities). On the other hand, we can think of several – observed or unobserved – sources of spatial variations in the popularity scores. An obvious candidate is the waste price: if residents hold local government responsible for the price of their waste disposal services relative to prices in neighbouring municipalities (see scenario 3a), popularity scores

Table 8.1 Descriptive statistics for the dependent variables

Variable	*Observations*	*Mean*	*S.D.*	*Min.*	*Max.*
Waste price 2006 [in c/l]	306	2.003	0.722	0.354	7.381
Waste price 2009 [in c/l]	306	2.263	0.667	0.417	5.000
Score local council 2006	306	6.672	0.378	5.000	7.600
Score local council 2009	306	6.727	0.422	5.300	7.700

Note: prices are in eurocents per litre and scores can theoretically range between 0 and 10.

might be spatially correlated. Other possible sources, such as regional political preferences, could have a similar impact on waste quantities.

Figure 8.1 shows waste prices and popularity scores, depicted on *choropleth* maps of Flanders: the shading of municipalities is proportional to the level of the variable of interest. The maps show that waste prices tend to be geographically clustered, that is, municipalities located in close proximity have similar levels of waste prices. The situation for popularity scores is less clear. Although the scores for some regions appear to be similar in proximate municipalities, overall spatial correlation seems rather low. However, there are several statistics that can be used to quantify a departure from spatial randomness. Table 8.2 reports the Moran's I statistic. This measure is a spatially weighted correlation coefficient, such that a positive value of this statistic points to spatial clustering, and a negative value is a sign of spatial dispersion. A spatial value of zero indicates the spatial pattern is completely random.[7] For all four variables we find a significant positive Moran's I statistic, indicating both waste prices and popularity scores are geographically clustered. Note that the statistic is relatively higher for waste prices compared to the statistic for the popularity scores, confirming the observations in the *choropleth* maps.

Recall that our model also predicted that the cost recovery rate for waste disposal and processing services could have an impact on popularity scores. We therefore include the difference between revenues generated through variable MSW pricing and the total cost for MSW collection and processing per kg as an additional variable when testing scenario 2. All cost and expenditure data are retrieved from public Flemish municipal accounts. Note that reported cost and revenue figures theoretically include all revenues and costs linked to the collection and processing of MSW, including recyclable waste fractions. Since the waste price captures only residual MSW prices (see below), the cost recovery rate used in our estimations has to be considered a proxy.

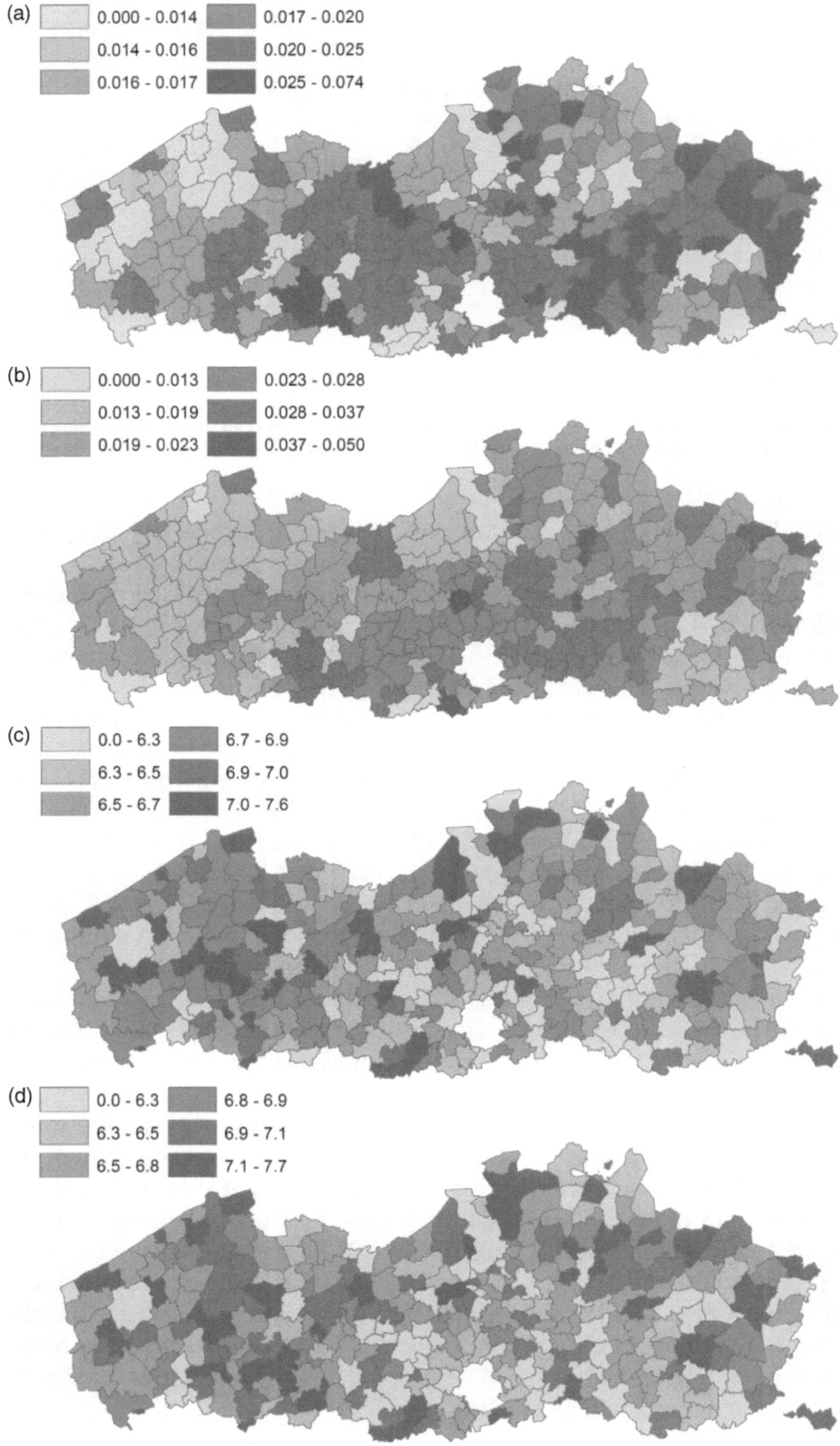

Figure 8.1 Geographical dispersal of waste prices in 2006 (a) and 2009 (b) and popularity scores in 2006 (c) and 2009 (d).

Table 8.2 Moran's I statistics[a]

Variable	*Statistic*	*P-value*
Waste price 2006	0.32	0.00
Waste price 2009	0.34	0.00
Score local council 2006	0.06	0.03
Score local council 2009	0.10	0.00

Note: [a] Obviously, computing the Moran's I statistic requires a matrix specifying the geographical relation between each observation. For a definition of this matrix, see section Stage 1 (price equation).

Although not explicitly modelled, revenue generated through variable waste pricing can also be used to finance other public services (e.g. public swimming pools, recreation facilities, cultural events, etc.). Since the provision of public goods can have a positive impact on popularity scores, we use total municipal expenditure per capita as an independent variable in the second scenario. Along with waste price and municipal finances, we include a vector of municipal characteristics that may influence popularity scores. Finally, we need a set of variables to instrument the price. We selected typical cost drivers such as a set of dummy variables indicating the type of waste processor used, a proxy for average processing costs and collection frequency.

Estimation strategy

Kinnaman and Fullerton (2000) indicate that policy choices in the context of municipal waste management are potentially endogenous in waste related models. Clearly the waste price is in this category. For example, if the decision to increase the price is correlated with unobserved characteristics that influence the popularity scores, it is possible that the estimates of policy impact are biased. To avoid this problem, we estimate waste prices in a first stage and, in a second stage, use predicted prices as an independent variable in our popularity functions. Although this allows us to circumvent the endogeneity problem, the use of two stage least squares (2SLS) has some particular data requirements. 2SLS estimates are generally inconsistent if the instruments are correlated with the error term in the equation of interest. In addition, so-called 'weak' instruments, that is, instruments that are poor predictors of the endogenous variable in the first-stage equation, can result in predicted values with very little variation, again resulting in biased estimates. Consequently, to avoid problems of price endogeneity, we need a set of instruments that, at least in part, can explain the variation in prices, but, conditional on the covariates, has no impact on waste generation. We expect that the municipal waste transport and processing cost drivers mentioned in the data section are the best candidates to meet these conditions.[8]

Stage 1: price equation

Estimating the price in a first stage imposes additional challenges. Controlling for all observable spatially correlated variables that influence waste prices might be insufficient to explain the observed geographic variability in prices. It is a well-established fact that ignoring any remaining spatial structure can lead to inconsistent estimates of the standard errors and, in the case of spatial interdependence among the dependent variables, to biased coefficient estimates (LeSage and Pace 2009). In our case, we can estimate our spatial models using the spatial maximum likelihood (ML) techniques developed by Anselin (1988). These spatial ML techniques allow for spatial correlation in the data to be dealt with in various ways. The most popular spatial model is the SAR or spatial autoregressive model, which includes a spatial lag of the dependent variable as one of the regressors. In our case such a lag could emerge in the equation as an additional independent variable consisting of a spatially weighted average price in neighbouring municipalities. Based on the empirical evidence in De Jaeger *et al.* (2009), the SAR model is likely to be the best fit for our data. In addition, including a spatial lag in the price equation allows us to test for spatial price interaction. To ensure that the SAR model is the best option, we use the robust Lagrange Multiplier (LM) tests for spatial autocorrelation and spatial lag dependency, developed by Anselin *et al.* (1996). As expected the test results that we discuss in the next section point to the spatial lag model as the most appropriate for our data:

$$p_w = \gamma_0 + \gamma_1 \bullet W \bullet p_w + R \bullet \theta' + v \qquad (12)$$

with v independent identically distributed ($v \sim \text{iid}(0, \sigma_v^2)$) and p_w, waste prices as before. $W \bullet p_w$ is the spatially weighted average of the prices in other municipalities and is derived from pre-multiplication of p_w with a spatial weight matrix. A general weight matrix W consists of NxN weight elements w_{ij}, where N is the number of municipalities. Each element w_{ij} measures the strength of the link between municipality *i* and municipality j. The diagonal elements of W are zero, that is, $w_{ij} = 0$ if i=j. The rows of the weighting matrix are normalized so that the elements w_{ij} sum to 1 in every row. We assume that spatial price interaction is mainly confined to neighbouring municipalities. Therefore, we assign to the elements w_{ij} of W the value 1 if municipalities *i* and j share a border, and 0 otherwise. This specific structure of the spatial weight matrix gives rise to the desired interpretation of $W \bullet p_w$ as the average price in neighbouring municipalities. If municipalities do mimic each other's prices, γ_1 should be positive.[9] Finally *R* is the set of cost drivers referred to above.

Table 8.3 Expected signs of coefficients for each scenario

Scenario 1	*Scenario 2*	*Scenario 3a*	*Scenario 3b*
$\alpha_1 < 0$	$\beta_1 > 0$	$\alpha_3 > 0$	$\beta_1 < 0$
	$\beta_2 > 0$		$\beta_2 < 0$
	$\beta_3 > 0$		$\beta_3 < 0$

Stage 2: popularity functions

In the second stage we estimate the popularity functions for each of the three scenarios, starting with scenarios 1 and 3. Econometric details for estimating scenario 2 are presented at the end of this section. Table 8.3 provides an overview of the expected signs of the coefficients in each scenario. Recall that in the first scenario consumers ignore local policymakers' budget constraints, which leads to a negative relation between popularity and MSW prices when we control for external effects. We therefore estimate the following model specification:

$$V = \alpha_0 + \alpha_1 \bullet \hat{p}_w + \alpha_2 \bullet y + \alpha_3 \bullet W \bullet \hat{p}_w + Z \bullet \varphi + \omega \quad (13)$$

where V is the popularity score for the entire council and $\hat{p}_w$ is the price predicted in first stage estimation results. As observations on the external effect are not available, α_1 should be significantly negative if the consumption effect is strong enough to offset the external effect (according to the first scenario). Note that we added the spatial lag of the predicted price ($W \bullet \hat{p}_w$) as one of the regressors in our estimation equation. This allows us to test whether the predictions from the model in scenario 3a hold. If consumers use the prices in neighbouring municipalities as a yardstick when judging their incumbents' pricing policy, α_3 should be significantly positive (i.e. higher average waste prices in neighbouring municipalities are advantageous for incumbents). Since the political economy literature states that incumbents are held responsible for economic progress (see, e.g., Vermeir and Heyndels, 2006), we include per capita income (y) as one of our explanatory variables. Obviously, we expect income to have a positive effect on popularity scores (i.e. α_2 is significantly positive). Finally Z consists of two additional municipal characteristics (unemployment rate and population density) which we use to proxy for economic welfare. Since general prosperity and general wellbeing probably promote higher levels of approval of the incumbents' policies, as in Vermeir and Heyndels (2006), we expect that the unemployment rate will negatively affect the incumbents' popularity scores. Population density is included to test for the effect of urbanization on popularity scores. We would expect residents living in densely populated areas to feel less sympathetic towards the incumbents and towards local policymakers in general, since poverty is often concentrated in large cities.

Recall that we suggested that consumers might also use historic waste prices in their own municipality as a yardstick (scenario 3b). To test this assumption, we relate the change in popularity scores ($\Delta V = V_{09} - V_{06}$) to the change in waste prices ($\Delta p_{w,t} = p_{w,t} - p_{w,t-1}$):

$$\Delta V = \beta_0 + \beta_1 \bullet \Delta p_{w,08} + \beta_3 \bullet \Delta p_{w.07} + \beta_4 \bullet \Delta y + \Delta Z \bullet \phi' + \tau \quad (14)$$

where $\Delta y = y_{09} - y_{06}$ and $\Delta Z = Z_{09} - Z_{06}$ are the changes in municipal characteristics (i.e. per capita income, cost recovery rate, per capita municipal expenditure, unemployment rate and population density) between 2006 and 2009.[10] Note that due to the lack of variance over time of our instruments for the price (i.e. weak instruments) we are not able to estimate the change in the prices in a first stage. Therefore we will rely on a single stage approach to test the predictions of this scenario. According to our model, any increase in the price should result in a decrease in the popularity score (i.e. β_1, β_2 and β_3 are negative), but it is more likely that more recent price changes have a higher impact in absolute values. Therefore, we distinguish specifically between the periods when changes occurred. Note the constant β_0 captures the overall difference in popularity scores between 2006 and 2009.

In order to test the predictions of the second scenario, where residents take account of local government budget constraints, we need to make an assumption. Recall the model in scenario 2 predicted that the impact of a change in the waste price on the popularity scores of incumbents will depend, among other things, on the consumer's relative level of waste generation $(\overline{w}_2 - w_2)$. Since we observe only average waste generation per capita at the municipal and not the individual level, we have no information on the distribution of waste generation within each municipality. However, waste generation data at the individual level are available for some other countries. For example, Tucker and Smith (1999) argue that measured waste generation rates appear to be quite well fitted by log-normal distributions. Also, data provided by an OECD household survey reveals that the distribution of waste generation (expressed as the number of waste bags presented) is strongly skewed to the right in nine out of the ten countries included in the survey (Kwan-Yim 2009). Although there seem to be some regional differences, data for all European countries in the sample reveal a similar right-tailed distribution. Therefore, we assume that average waste levels are higher than the median waste level in most Flemish municipalities. In other words, if we draw a random sample in each municipality, most respondents will produce less waste than the average produced per capita in their municipality. Recall our model predicted that if $\overline{w}_2 > w_2 (\overline{w}_2 < w_2)$ an increase in the price for waste disposal would have a positive (negative) impact on popularity scores when we control for the cost of recovery rate and the external effects of waste collection and processing. Therefore, if scenario 2 holds, and assuming the distribution of waste generation is skewed to the right, the majority of respondents will prefer an increase to a decrease in the price for waste disposal.

However, if the popularity score of incumbents decreases linearly in the utility loss of the respondents, the average popularity score will be zero. To show this, note first that the model derived in the previous section states that utility depends linearly on the difference between the consumer's waste level and the average waste level. Second, for a randomly drawn group of respondents within a municipality, the expected average distance from the mean waste level will be zero, even if the distribution for waste generation is strongly skewed. Therefore, average utility loss due to a change in price will also be zero. Because we only have the average popularity score for each municipality, we cannot test the predictions of the model in scenario 2 about linear popularity scores. However, we believe that respondents' approval of the pricing policy will be related mainly to the sign of the impact on their utility and less to the magnitude of that impact. It is likely that residents react more to the fact that the incumbent's policy affects their utility, than to the total utility change. This assumption implies marginal popularity $V'(u_i) > 0$ and $V''(u_i) < 0$. Therefore, if scenario 2 holds, we can expect average popularity scores to increase as the price increases. To test the predictions of this model specification, we can employ the estimation results for the other scenarios. If an increase in prices results in an increase in popularity scores, β_1, β_2 and β_3 should be significantly positive. Note that this is the reverse of what we expect in scenario 3b (see Table 8.3).

Before discussing the results, were need to refer to some econometric issues. First, Figure 8.1 and Moran's I statistics in Table 8.2 reveal a moderate degree of spatial clustering of the popularity scores. Controlling for all observable spatially correlated variables which influence the popularity scores might still not be sufficient to explain the geographic variability of the popularity scores. As already argued, ignoring the spatial structure can lead to biased coefficient estimates and inconsistent estimates of the standard errors. Therefore, we draw on the LM tests for spatial autocorrelation and spatial lag dependence, to detect spatial autocorrelation after controlling for observable covariates (in our case the regressors). As the discussion of results in the next section shows, spatial correlation is no longer present when estimating the above model specification. Second, we cannot use γ_1 to calculate the fitted values of the waste price since the variable $W{\bullet}p$ is not exogenous. We included this variable in the first stage because otherwise the other coefficients could suffer from omitted variable bias. Finally, since in our popularity function we use fitted values for the price, the covariance matrix of the estimator in this step includes some noise induced by the first-stage estimates. Therefore, we use the limited information ML procedure to correct the covariance matrix. Thus, the final estimation procedure involves four steps (except for scenario 3b): (1) estimating the price equation using the spatial MLL method in Anselin (1988); (2) using only the estimated coefficients of the exogenous variables to calculate the fitted value $\hat{p}_w$ of the price variable p_w. Note that estimation of

Table 8.4 Model specification tests

	Lagrange multiplier		*Robust Lagrange multiplier*	
	Statistic	*P-value*	*Statistic*	*P-value*
2006 waste price	16.678	0.000	5.787	0.016
2009 waste price	7.735	0.005	8.105	0.004
Change price	9.74	0.002	10.043	0.002
2006 popularity score	0.042	0.837	0.131	0.717
2009 popularity score	2.298	0.130	0.006	0.937
Change popularity score	0.150	0.699	0.042	0.837

Notes: We also tested the spatial error and spatial Durbin model. The LM tests for spatial autocorrelation and the spatial Hausman test reveal that both model specifications are not supported by our data. For more details on the tested models, see Anselin (1988). A detailed description of the model specification tests can be found in Anselin *et al.* (1996) (LM tests for spatial autocorrelation) and LeSage and Pace (2009) (spatial Hausman test).

our model predictions in scenario 2 and 3b also requires separate predictions for year by year changes in the price. Regressors in the latter case are simply yearly changes to the independent variables used to predict the prices in levels; (3) estimating the popularity function replacing the price variable p_w by its fitted value $\hat{p}_w$; (4) using the limited information ML procedure to correct the covariance matrix.

By following this procedure we circumvent the endogeneity problem, while still retrieving the information on tax mimicking and popularity functions across municipalities.

Estimation results

In this section we present and comment on the estimation results. We start by briefly discussing the model specification statistics (Table 8.4), then discuss the first stage regression results (Table 8.5) and the estimation results for the popularity scores (Table 8.6). As mentioned in the previous section, we use LM tests to check whether the above spatial model specifications fit our data. For the price equation both the robust and non-robust LM statistics

Table 8.5 Estimation results for the waste price

Variable	*ML 2006*	*ML LAG 2006*	*ML 2009*	*ML LAG 2009*
Wp	–	0.290***	–	0.204***
	–	(0.073)	–	(0.075)
Unemployment rate	−0.051*	−0.050**	−0.009	−0.008
[in %]	(0.027)	(0.025)	(0.023)	(0.022)
Average income	0.018	0.012	0.024	0.018
[in €1,000]	(0.018)	(0.017)	(0.014)	(0.014)
Constant	1.774***	1.316**	1.589***	1.303***
	(0.571)	(0.549)	(0.473)	(0.463)
Cost drivers				
Proxy total cost	3.456***	3.198***	−0.301	−0.505
[in €/kg]	(1.096)	(1.031)	(0.903)	(0.866)
Collection frequency	−0.011***	−0.008**	−0.009***	−0.008**
[in collections/ year]	(0.004)	(0.004)	(0.003)	(0.003)
Processor 1	−0.054	−0.059	0.090	0.079
	(0.149)	(0.140)	(0.130)	(0.124)
Processor 2	−0.344***	−0.276**	−0.091	−0.070
	(0.117)	(0.111)	(0.100)	(0.095)
Processor 3	−0.209	−0.177	−0.331**	−0.302**
	(0.166)	(0.156)	(0.140)	(0.134)
Processor 4	0.082	0.072	0.305	0.287
	(0.223)	(0.209)	(0.190)	(0.181)
Processor 5	0.134	0.109	0.082	0.063
	(0.162)	(0.153)	(0.141)	(0.134)
Processor 6	0.130	0.068	0.324**	0.284**
	(0.173)	(0.163)	(0.150)	(0.144)
Processor 7	0.139	0.041	0.430***	0.349***
	(0.122)	(0.118)	(0.107)	(0.106)
Processor 8	−0.894***	−0.651***	−0.638***	−0.550***
	(0.239)	(0.232)	(0.206)	(0.200)
Processor 9	0.138	0.085	0.288***	0.254***
	(0.096)	(0.092)	(0.083)	(0.081)
Processor 10	0.301*	0.227	0.614***	0.549***
	(0.165)	(0.156)	(0.141)	(0.137)
Processor 11	0.395**	0.308*	0.717***	0.654***
	(0.184)	(0.174)	(0.159)	(0.154)
Processor 12	0.136	0.088	0.582***	0.491***

Table 8.5 (cont.)

Variable	*ML 2006*	*ML LAG 2006*	*ML 2009*	*ML LAG 2009*
	(0.175)	(0.165)	(0.153)	(0.150)
Processor 13	0.662***	0.542***	0.804***	0.698***
	(0.169)	(0.162)	(0.147)	(0.145)
Log likelihood	−178.690	−279.750	−185.584	−240.590
LR test of rho=0	–	16.678***	–	7.735***

Note: robust standard errors between brackets.

Table 8.6 Estimation results for the popularity scores

Variable	*ML -2006*	*ML -2009*	*ML-change*
Price [in c/l]	−0.107***	−0.068*	–
	(0.032)	(0.041)	–
Wp	−0.157	−0.133	–
	(0.106)	(0.115)	–
Price change 2009 [in c/l]	–	–	−0.093**
	–	–	(0.043)
Price change 2008 [in c/l]	–	–	−0.095*
	–	–	(0.054)
Price change 2007 [in c/l]	–	–	−0.023
	–	–	(0.054)
Cost recovery rate [in €/kg]	–	–	0.516
	–	–	(0.756)
Public goods [in €1,000/cap]	–	–	0.120
	–	–	(0.129)
Unemployment rate [in %]	−0.026**	−0.055***	−0.017
	(0.013)	(0.017)	(0.030)
Average income [in €1,000]	0.003	−0.007	−0.026
	(0.008)	(0.010)	(0.037)
Population density [in pop/km²]	−0.175***	−0.115**	−3.171
	(0.052)	(0.062)	(1.955)
Constant	7.298***	7.681***	0.097***
	(0.260)	(0.277)	(0.046)
Observations	306	306	306
Log likelihood	−114.318	−149.954	−73.496

Notes: robust standard errors between brackets; socio-demographic variables are in changes (2009–2006) for the last column. Price variables are instrumented in columns 'ML-2006' and 'ML-2009'.

for the presence of a lagged dependent variable are significantly positive. The model with a spatially lagged dependent variable is preferred over a model without a spatial lag. For the popularity score equations we have no reason to assume that popularity scores depend endogenously on the popularity scores in other geographically close municipalities, but spatial variation could have other sources (see data section). Following the LM test statistics, however, we find no evidence of spatial correlation, indicating that the model specification proposed in the previous section is the most relevant for our data set.

Price equation

Table 8.5 shows the results of the price equation, the first step in our estimation procedure. We display the results without the spatial lag of the dependent price variable, estimated by ML in column 1 for 2006 (denoted ML 2006) and in column 3 for 2009 (denoted ML 2009); and with the spatial lag of the price variable, estimated by spatial ML in column 2 for 2006 (denoted ML LAG 2006) and in column 4 for 2009 (denoted ML LAG 2009).

The spatial lag of the price variable Wp is significant in all the relevant equations. For 2006 the value of the spatial lag parameter is 0.29. This implies that when the average waste price in the neighbouring of municipality i increases by one eurocent, the price in municipality i increases by 0.29 eurocents. The coefficient for 2009 is lower, but still significant at the 1 per cent level. The value of approximately 0.23 suggests a municipality counters a change of one eurocent between 2006 and 2009, by a change of 0.23 eurocents. These results suggest municipalities mimic price level changes in neighbouring municipalities.

Coefficients and inferences for the other variables appear to be quite similar between the LAG and non-LAG estimates. However, the observation year seems to make a difference. The unemployment rate has the expected significant negative sign for 2006, but for the 2009 estimate the effect is no longer significant. Contrary to what we expected, average income does not significantly explain price variations. Apparently local policymakers do not take account of the financial endowments of their residents when deciding about waste prices. The average cost variable is significantly positive for the 2006 estimate, indicating that municipalities tend to charge their residents according to the costs of disposal and processing of waste. The coefficients for 2009 are no longer significant. Collection frequency seems to have a negative impact on prices. This might indicate that municipalities benefit from increasing returns to scale. However, it is possible also waste bags might be less full because of more frequent waste collection, which in turn could lead to lower waste bag prices. The effect of the waste processor also differs between observation years. Only processors eight and 13 have the same significant effect on prices in both years.

Table 8.7 Timing changes in pricing policy

Years before election	*% municipalities price increase*	*% municipalities*
Price decrease		
0	3.92	0.33
1	5.88	0.00
2	2.29	0.98
3	19.28	2.29
4	16.01	1.31
5	11.11	5.23

Popularity scores

We next look at the results of the popularity equations (Table 8.6). Columns 1 and 2 display the ML estimation results for 2006 and 2009 (two stage approach) while column 3 displays the ML results for changes in popularity scores (single stage approach).

Based on our theoretical model, we are interested mainly in the impact of the price variables on the popularity scores. If we look at the estimates in levels (columns 1 and 2), we can see that prices have a significantly negative impact on the incumbent's popularity. Note the spatial lag of the price variable Wp is not significant in any of the relevant equations. This implies that the hypothesis that consumers use the price in a neighbouring municipality as a yardstick for judging their own pricing schedule is not supported by our data. The coefficient estimates for year by year changes in prices (column 3) reveal that only the most recent changes in prices have an impact on popularity scores. Apparently consumers have short memories and hold incumbents responsible only for recent policy actions. It is interesting that local policymakers seem not to take this into account. If we look at the timing of increases in the waste price, we see that most municipalities changed the price in the three years after an election (see Table 8.7). We can conclude that scenarios 1 and 3b are supported by the data. Consumers seem to look at both absolute levels and recent changes in prices when judging incumbents. The price levels in neighbouring municipalities and policymakers' budget constraints appear to be less important.

As in the price equations, average income seems to have no impact on the dependent variable. Possibly consumers realize that economic policy is a mainly federal and regional responsibility and municipal policymakers do not have much influence. The coefficients of the unemployment rate and population density, however, have the expected negative sign, although, in the ML change estimate, the effect for unemployment rate is not significant. The latter result could be the consequence of a general negative attitude towards incumbents among consumers in more straitened circumstances.

Concluding remarks

This chapter proposed a simple theoretical model of consumers' preference regarding local municipalities' MSW pricing policy choices. The model predictions are tested on 306 (out of 308) Flemish municipalities, for 2006 and 2009. The results indicate that residents hold local policymakers responsible for residual waste prices, but do so without comparing against prices charged in neighbouring municipalities. Political costs in terms of popularity scores seem to depend on absolute price levels and changes in price levels.

These findings imply that the observed waste price mimicking among municipalities has no advantage in terms of re-election chances. So what is the motive for basing waste prices on the prices in neighbouring municipalities? Other sources of waste price mimicking, such as tax competition, seem unlikely since it is doubtful that residents would move to another municipality based on the price charged for waste disposal. De Jaeger *et al.* (2009) tested whether waste exports (dumping) could lead to spatial correlations in price levels. They argue that a much lower waste price in a neighbouring municipality could induce 'waste tourism' (waste export), involving the residents in the more expensive municipality transporting their waste for collection in an adjacent (cheaper) municipality. If local decision makers set the waste prices to be similar or very close to the price in neighbouring municipalities to prevent this behaviour, waste prices will be spatially correlated. However, they found no evidence of waste streams driven by differences in price levels, indicating local policymakers probably do not consider this issue when deciding on their waste pricing scheme.

It could be argued that, as long as politicians believe voters will try to overcome information asymmetries by comparing local policies with neighbouring municipalities, policymakers will react accordingly. This would be sufficient to explain the upward sloping tax reaction functions in the data. Ashworth and Heyndels (1997), for example, use a sample of Flemish politicians to investigate politicians' opinions about local property and income taxes. They found evidence that tax policy in neighbouring jurisdictions affects the perceived political cost of the focal jurisdiction's tax rate. There is other empirical evidence showing that, in Belgium, local property and income taxes are strongly correlated among neighbouring municipalities (Heyndels and Vuchelen 1998). Note that, in this case, the politicians' perceptions would seem to be justified, since research by Vermeir and Heyndels (2006) reveals that electoral punishment depends on the level of the property and income taxes in neighbouring municipalities. Given the evidence of yardstick voting in the case of property and income taxes, it seems plausible that local policymakers would believe also that the political cost of waste prices depends on price levels in neighbouring municipalities, inducing them to engage in price mimicking. For more conclusive results, we need more data and further research.

Appendix: Descriptive statistics

Table 8.A.1 Descriptive statistics independent variables

Variable	*Mean*	*Standard Deviation*	*Min*	*Max*
Unemployment rate 2006 [in %]	6.413	2.004	3.260	15.370
Unemployment rate 2009 [in %]	5.748	1.835	2.780	14.600
Average income 2006 [in €1,000]	26.795	3.135	19.652	38.522
Average income 2009 [in €1,000]	27.652	3.352	20.363	39.910
Proxy total cost 2006 [in €/kg]	0.119	0.041	0.000	0.240
Proxy total cost 2009 [in €/kg]	0.122	0.044	0.000	0.260
Collection frequency 2006 [in collections/year]	37	13.748	2	99
Collection frequency 2009 [in collections/year]	36	13.694	6	100
Cost recovery rate 2006 [in €/kg]	−0.097	0.036	−0.251	0.000
Cost recovery rate 2009 [in €/kg]	−0.098	0.042	−0.234	0.000
Public goods 2006 [in €1000/cap]	0.783	0.227	0.421	2.129
Public goods 2009 [in €1000/cap]	0.855	0.286	0.000	2.376
Population density 2006 [in pop/km^2]	0.524	0.448	0.053	3.138
Population density 2009 [in pop/km^2]	0.530	0.451	0.053	3.137

Note: The average income in 2009, the proxy for the total cost in 2009 and the expenditures for public goods in 2009 are partly based on provisional data as not all observations have undergone the final check by the responsible authorities.

Notes

1 An overview of the empirical evidence can be found in Vermeir and Heyndels (2006).
2 For later reference we define the total volume of waste W as $\overline{w} \cdot n$, where $\overline{w}$ is the average amount of waste generated per resident and n is the number of residents in the municipality.
3 Note that we consider only the interior solution since we assume $\lim_{z \to 1} c'(z) = +\infty$.
4 Residual MSW consists of: (1) all residual waste presented by households at the kerbside; (2) bulky household refuse; and (3) municipal waste. Note that solid waste generated by companies, schools, hospitals, prisons, etc., is collected by private waste collection and processing firms and is not included in the definition of residential solid waste (OVAM 2002).
5 The results of the questionnaire are available at: www.nieuwsblad.be/burgemeester [in Dutch].
6 In order to avoid bias arising from an over- or under-representation of certain population related characteristics, a proportionally interlaced stratified sample of the population is used.
7 For more details on Moran's I statistic see Anselin and Kelejian (1997).

8 Various diagnostic tests, such as the Sargan-Hansen test for overidentifying restrictions (Sargan 1958; Hansen 1982) and weak identification tests (Stock and Yogo 2005), allow us to verify whether these conditions are met. The test results show our instruments are exogenous, but the instruments are not always able to explain the variations in the price in a sufficient way.

9 We could use other definitions of geographical proximity, such as the inverse of the distance between municipalities i and j or the inverse of the distance squared. We use these weight matrices as a robustness check, but chose the weight matrix W because it provides the best explanation for spatial clustering in the disturbances (highest log likelihood value).

10 The literature provides evidence that, in some circumstances, politicians adapt their policies according to their popularity stock (Frey and Schneider 1978; Schneider and Pommerehne 1980). If incumbents with high popularity scores are more likely to increase the waste price, estimates based on the exact level of the popularity score may be biased. Therefore, we estimate the entire equation in changes.

References

Anselin, L. (1988) *Spatial Econometrics: Methods and Models*, Boston: Kluwer Academic.

Anselin, L. and Kelejian, H.H. (1997) 'Testing for Spatial Error Autocorrelation in the Presence of Endogenous Regressors', *International Regional Science Review*, 20: 153–82.

Anselin, L., Bera, A.K., Florax, R. and Yoon, M.J. (1996) 'Simple Diagnostic Tests for Spatial Dependence', *Regional Science and Urban Economics*, 26: 77–104.

Ashworth, J. and Heyndels, B. (1997) 'Politicians' Preferences on Local Tax Rates: An Empirical Analysis', *European Journal of Political Economy*, 13: 479–502.

Ashworth, J., Benny, G. and Heyndels, B. (2006) 'Determinants of Tax Innovation: The Case of Environmental Taxes in Flemish Municipalities', *European Journal of Political Economy*, 22: 223–47.

Besley, T.J. and Case, A.C. (1995) 'Incumbent Behavior: Vote Seeking, Tax Setting and Yardstick Competition', *American Economic Review*, 85: 25–45.

Bordignon, M., Cerniglia, F. and Revelli, F. (2003) 'In Search of Yardstick Competition: A Spatial Analysis of Italian Municipality Property Tax Setting', *Journal of Urban Economics*, 54: 199–217.

Bordignon, M., Cerniglia, F. and Revelli, F. (2004) 'Yardstick Competition in Intergovernmental Relationships: Theory and Empirical Predictions', *Economics Letters*, 83: 325–33.

Bosch, N. and Solé-Ollé, A. (2007) 'Yardstick Competition and the Political Costs of Raising Taxes: An Empirical Analysis of Spanish Municipalities', *International Tax and Public Finance*, 14: 71–92.

Brueckner, J.K. (2003) 'Strategic Interaction Among Governments: An Overview of Empirical Studies', *International Regional Science Review*, 26: 175–88.

Congleton, R. (2003) 'The Median Voter Model', in C.K. Rowley and F. Schneider (eds), *The Encyclopedia of Public Choice*, Dordrecht: Kluwer Academic Press, 382–386.

De Jaeger, S., Eyckmans, J., Van Parys, S. and Verbeke, T. (2009) 'Waste Demand in the Context of Waste-price Mimicking and Waste Tourism: A Case Study of Flemish Municipalities', *mimeo*.

Downs, A. (1957) *An Economic Theory of Democracy*, New York: Harper & Row.
Eshet, T., Ayalon, O. and Shechter, M. (2006) 'Valuation of Externalities of Selected Waste Management Alternatives: A Comparative Review and Analysis', *Resources, Conservation and Recycling*, 46: 335–64.
Frey, B. S. and Schneider, F. (1978) 'A Politico-economic Model of the United Kingdom', *The Economic Journal*, 88: 243–53.
Fullerton, D. and Kinnaman, T.C. (1995) 'Garbage, Recycling and Illicit Burning or Dumping', *Journal of Environmental Economics and Management*, 29: 78–91.
Geys, B. and Vermeir, J. (2008) 'The Political Cost of Taxation: New Evidence from German Popularity Ratings', *Electoral Studies*, 27: 633–48.
Hansen, L. (1982) 'Large Sample Properties of Generalized Method of Moments Estimators', *Econometrica*, 50: 1029–54.
Heyndels, B. and Vuchelen, J. (1998) 'Tax Mimicking among Belgian Municipalities', *National Tax Journal*, 51: 89–101.
Kiel, K. and McClain, K. (1995) 'House Prices during Siting Decision Stages: The Case of an Incinerator from Rumor through Operation', *Journal of Environmental Management*, 28: 241–55.
Kinnaman, T.C. and Fullerton, D. (2000) 'Garbage and Recycling with Endogenous Local Policy', *Journal of Urban Economics*, 48: 419–42.
Kwan-Yim, K. (2009) 'Household Behavior and Environmental Policy: Waste Generation', paper presented at the OECD Conference on Household Behaviour and Environmental Policy, Paris: OECD.
LeSage, J.P. and Pace, R.K. (2009) *Introduction to Spatial Econometrics*, New York: CRC Press.
OVAM (Openbare Vlaamse Afvalstoffenmaatschappij) (2002) 'Uitvoeringsplan Huishoudelijke Afvalstoffen 2003–2007', Mechelen: OVAM.
Powell, G.B. and Whitten, G.D. (1993) 'A Cross-National Analysis of Economic Voting: Taking Account of the Political Context', *American Journal of Political Science*, 37: 391–414.
Revelli, F. (2001) 'Spatial Patterns in Local Taxation: Tax Mimicking or Error Mimicking?', *Applied Economics*, 33: 1101–7.
Revelli, F. (2002) 'Local Taxes, National Politics and Spatial Interactions in English District Election Results', *European Journal of Political Economy*, 18: 281–99.
Sargan, J.D. (1958) 'The Estimation of Economic Relationships Using Instrumental Variables', *Econometrica*, 26: 393–415.
Schneider, F. and Pommerehne, W.W. (1980) 'Politicoeconomic Interactions in Australia: Some Empirical Evidence', *The Economic Record*, 56: 113–31.
Solé-Ollé, A. (2003) 'Electoral Accountability and Tax Mimicking: The Effects of Electoral Margins, Coalition Government, and Ideology', *European Journal of Political Economy*, 19: 685–713.
Stock, J.H. and Yogo, M. (2005) 'Testing for Weak Instruments in Linear IV Regression', in D.W. Andrews and J.H. Stock (eds), *Identification and Inference for Econometric Models: Essays in Honor of Thomas Rothenberg*, Cambridge: Cambridge University Press, pp. 80–108.
Tresch, R. (2002) *Public Finance – A Normative Theory*, San Diego: Academic Press.

Tucker, P. and Smith, D. (1999) 'Simulating Household Waste Management Behaviour', *Journal of Artificial Societies and Social Simulation*, 2.

Vermeir, J. and Heyndels, B. (2006) 'Tax Policy and Yardstick Voting in Flemish Municipal Elections', *Applied Economics*, 38: 2285–98.

Wilson, J.D. (1996) 'Theories of Tax Competition', *National Tax Journal*, 52: 269–304.

Index